Ionic Liquids-based Nanofluids

Ionic Liquids-based Nanofluids provides a comprehensive exploration of nanoparticle dispersions in ionic liquids, covering their design, physicochemical behaviour, and technological relevance. It builds from the foundational principles of colloidal science and ionic liquid chemistry to advances in synthesis strategies, interfacial phenomena, and structure–property relationships.

Integrating theoretical frameworks with experimental insights and application-driven perspectives, the book highlights the transformative potential of ionic liquids-based nanofluids in diverse sectors, including energy harvesting and storage systems, advanced thermal management, and electrochemical devices. It addresses sustainability dimensions, including toxicity assessment, environmental implications, lifecycle considerations, and safe handling practices, emphasizing responsible development and deployment.

This book will interest industry researchers and graduate students studying nanofluid applications within the fields of mechanical, materials, chemical, and environmental engineering.

Ionic Liquids-based Nanofluids

Edited by
Aswathy Joseph, S M Sohel Murshed,
and Suresh Mathew

CRC Press
Taylor & Francis Group
Boca Raton London New York

CRC Press is an imprint of the
Taylor & Francis Group, an **informa** business

Designed cover image: Shutterstock

First edition published 2027
by CRC Press
2385 NW Executive Center Drive, Suite 320, Boca Raton FL 33431

and by CRC Press
4 Park Square, Milton Park, Abingdon, Oxon, OX14 4RN

CRC Press is an imprint of Taylor & Francis Group, LLC

ISBN: 978-1-032-97753-9 (hbk)
ISBN: 978-1-032-97755-3 (pbk)
ISBN: 978-1-040-65465-1 (ebk)

DOI: 10.1201/9781003595250

Typeset in Times LT Std
by KnowledgeWorks Global Ltd.

Contents

About the Editors

Aswathy Joseph is Assistant Professor at the Postgraduate and Research Department of Chemistry, Catholicate College, Kerala, India. She received a Bachelor's degree (2008) and Master's degree (2010) in Chemistry from the University of Kerala, India. She received a PhD (2019) from Mahatma Gandhi University (MGU), India and a Postgraduate Diploma in Education Technology and Management (2021) from ICAR-NAARM, UoH, India. She was awarded the Kerala University Merit Scholarship (2008) and Jawaharlal Nehru Scholarship for Doctoral Studies (2015–2017) from Jawaharlal Nehru Memorial Fund, New Delhi. She also qualified for the UGC-CSIR National Eligibility Test for Lectureship in Chemical Science in December 2010. She has numerous publications in reputed international journals, including two major reviews and one edited book. Her areas of research interest are ionic liquids, ferrofluids, ionanofluids, polymer-ionic liquid thin films, and magneto-luminescent nanomaterials.

S M Sohel Murshed received his PhD in Mechanical and Aerospace Engineering from Nanyang Technological University, Singapore. He is currently Professor at the Department of Mechanical Engineering of the Instituto Superior Técnico, University of Lisbon, Portugal. He was Visiting Professor at Rochester Institute of Technology (RIT), USA (2021–2022) and Indian Institute of Technology (IIT Dhanbad), India (2022). Prof. Murshed is listed in the rankings of World's Top 2% Scientists (both career and single year of 2021 impact lists) that was published by Stanford University in October 2023. In addition to 12 books and 35 book chapters, Prof. Murshed has authored/co-authored more than 240 scientific publications in leading international journals and conferences. His current citation count is 11,100 with h-index of 43 (Google Scholar). He is a member of the American Society of Mechanical Engineers (ASME), American Society of Thermal and Fluids Engineers (ASTFE), and European Thermoelectric Society (ETS). He has regularly served as an expert evaluator for the EC and for funding agencies of different countries across the globe.

Suresh Mathew is Alexander-von-Humboldt Fellow and Senior Professor of Inorganic Chemistry at Mahatma Gandhi University, India. He is Founder and Director of the Advanced Molecular Materials Research Centre (AMMRC) at Mahatma Gandhi University. He completed his post-graduation from CUSAT, India in Applied Chemistry and earned a Ph.D. in Chemistry from the University of Kerala, India in Collaboration with Vikram Sarabhai Space Centre, Trivandrum. He was Visiting Fellow at Jawaharlal Nehru Centre for Advanced Scientific Research (JNCASR), Bangalore (2000); Lunghwa University of Science and Technology, Taiwan (May to July 2009); and Rzeszow University of Technology, Poland (November 2015 and February 2017). Dr. Mathew has participated in the Leadership for Academician Programme at IIT Bombay and NTU Singapore. He has published 230 research papers in international journals and conferences, three patents, three edited books, and numerous book chapters.

Preface

The increasing demand for energy-efficient and environmentally sustainable thermal management technologies has intensified interest in advanced fluids that surpass the limitations of conventional heat transfer media. Among these, ionic liquid-based nanofluids, also known as ionanocolloids, have emerged as a distinctive class of multifunctional nanomaterial-based fluids that integrate principles of thermal science, chemistry, and nanotechnology. This book provides a concise yet comprehensive reference for researchers, engineers, and graduate students in energy, materials science, and chemical engineering, encompassing fundamental concepts, recent progress, and emerging perspectives on ionanocolloids and related systems.

Chapter 1 presents the idea of ionanocolloids and describes how they have developed from curiosities at the laboratory scale to viable options for cutting-edge thermal and electrochemical processes. It highlights their unique characteristics, the evolution of their synthesis, and the mechanistic processes that control their functionality.

Ionanofluids' synthesis methods, electrochemical behaviour, and energy-related uses are covered in Chapter 2, with particular attention paid to their potential in dye-sensitized solar cells, redox flow batteries, and desulfurization systems.

In Chapter 3, ionanocolloid electrolytes for electrochemical storage are discussed, along with how nanoscale alterations improve ionic transport, interface stability, and energy density in systems based on lithium and sodium.

Ionanocolloids' thermophysical characteristics are examined in Chapter 4, with a focus on important variables like density, specific heat, thermal conductivity, and stability. It combines theoretical understanding and experimental data to direct performance optimization.

Chapter 5 explores the mechanisms of heat transfer in ionanocolloids by examining molecular-level thermal transport, interfacial nanolayer formation, and nanoparticle clustering from both an experimental and computational perspective.

In conclusion, Chapter 6 critically analyses the opportunities and difficulties that ionanocolloids face in real-world applications, pointing out deficiencies in lifecycle analysis, modelling, characterization, and scalability. It also describes tactical ways to turn these cutting-edge materials from theoretical ideas into sustainable, commercially feasible technologies.

The aim of this book is to serve as a reference and a source of scholarly inspiration for future education and research on the development and application of ionanofluids. By integrating theoretical modelling, experimental methodologies, and sustainability frameworks, it aspires to foster the rational development of high-performance, environmentally responsible thermofluids and energy materials. We hope that this collective effort will foster interdisciplinary collaboration and contribute meaningfully to the global pursuit of next-generation thermal management and energy technologies.

Contributors

Aruna Kumar Barick
Department of Chemistry
Veer Surendra Sai University of Technology
India

Ajaya Kumar Behera
P.G. Department of Chemistry
Utkal University
India

Shreelata Behera
Department of Chemistry
Veer Surendra Sai University of Technology
India

Bigyan Ranjan Jali
Department of Chemistry
Veer Surendra Sai University of Technology
India

Aswathy Joseph
Department of Chemistry
Catholicate College
India

Suresh Mathew
School of Chemical Sciences
Mahatma Gandhi University
India

Annada Mishra
Department of Chemistry
Veer Surendra Sai University of Technology
India

Pooja Mohapatra
Department of Chemistry
Veer Surendra Sai University of Technology
India

S M Sohel Murshed
IDMEC, Department of Mechanical Engineering
Instituto Superior Técnico, University of Lisbon
Portugal

Bibhu Prasad Sahoo
School of Applied Sciences (Chemistry)
KIIT Deemed to be University
India

Saleja Sahoo
Department of Chemistry
Veer Surendra Sai University of Technology
India

Lipsa Shubhadarshinee
Department of Chemistry
Veer Surendra Sai University of Technology
India

P. S. Rama Sreekanth
School of Mechanical Engineering
VIT-AP University of Technology
India

1 An Introduction to Ionanocolloids

Aswathy Joseph, S M Sohel Murshed, and Suresh Mathew

1.1 INTRODUCTION

1.1.1 Background, Definition, and Nomenclature

Ionic liquid (IL)–based nanofluids or *ionanocolloids (INCs)*, also sometimes called *nanoparticle-enhanced ionic liquids (NEILs)*, represent a specialized subclass of nanofluids. These are a novel class of heat transfer fluids that incorporate the distinctive properties of ILs and nanoparticles. These nanofluids are obtained by stabilizing nanoscale materials (typically 1–100 nm), such as metals, oxides, carbons, polymers, and metal–organic frameworks (MOFs), in an IL base fluid. Thus, created colloidal suspensions demonstrate advanced thermophysical properties compared to their basic constituents. It is important to acknowledge the inconsistent nomenclature of this nanofluid subclass observed in the literature. In the context of advanced heat transfer fluids and nanomaterial dispersions, the terms *ionanofluid* and "*ionanocolloid*" are being used more frequently. Although the terms are occasionally used synonymously, an increasing amount of research aims to make these distinctions clear for more accurate scientific communication and application development. From a chemical engineering and colloidal and thermal science point of view, the term "*ionanocolloids*", which was coined by the group of Murshed, is scientifically appropriate [1]. Unlike conventional nanofluids, where nanoparticles are suspended in molecular solvents (water, glycols, oils), *INCs* exploit the unique physicochemical environment of ILs, i.e., the high charge density, negligible vapor pressure, and tunable polarity, to yield dispersions that are often more thermally stable, non-volatile, and electrochemically functional. In this chapter, we adopt the term *INC* to mean a *colloidal dispersion of nanoparticles within an ionic liquid medium*, where the IL serves both as the continuous phase and as an active interfacial stabilizer. The nanoparticles may be metallic, metal oxide, carbon-based, or polymeric and are stabilized primarily through strong ion–particle interactions, electrostatic layering, and solvation forces intrinsic to the IL structure.

Meanwhile, demand for novel fluids that could operate under extreme thermal conditions, outperforming conventional heat transfer fluids, was thriving. This has successfully paved the way for new investigations on IL-based nanofluids. Contemporaneously, the coming out of novel ILs was also on the research limelight. ILs, by their very nature, are salts, composed of organic cations and organic or inorganic anions. Most of them have melting points below 100°C, a wide liquid range, low vapor pressure, good thermo-oxidative stability, and an ability to defy flammability [2].

DOI: 10.1201/9781003595250-1

Most of these traits are absent in water- or oil-based nanofluids. Besides, the "designer" nature of ILs made property tailoring more feasible by modifying the cation–anion pair. This further enhanced its appeal to a great extent. Certain imidazolium- or pyrrolidinium-based ILs can operate at overwhelmingly high temperatures, even up to 350°C. These properties make ILs viable alternatives of current commercial high-temperature base fluids, such as biphenyl or diphenyl oxide, that degrade above 200°C [3]. The excellent thermal stability of IL and the improved thermal conductivity and heat capacity of the nanofluid idea work in coherence to support the development of novel INCs [4]. This collaboration seeks to develop a better medium for uses, including concentrated solar power (CSP) and sophisticated electronic cooling systems that require high thermal stability and exceptional heat transfer performance. This chapter aims to provide a brief introduction to INCs, outlining their advances in various fields and providing the further direction of research and development.

1.1.2 Chronology and Evolution of Ionanocolloid Research

The groundwork on INCs progresses from two separate research streams: nanotechnology and IL chemistry. The field of nanofluids itself has roots in concepts developed in the 1960s by Stephen Papell's creation of ferrofluids [5]. However, it was not until the early 2000s that ILs became the focus of serious engineering efforts as they were shown to have profound chemical and physical advantages [6]. The research on INCs, as we know them today, started when researchers prepared them, adding different nanoparticles, such as gold (Au), copper oxide (CuO), aluminum oxide (Al_2O_3), and multi-walled carbon nanotubes (CNTs) (MWCNTs) to imidazolium-based ILs [7, 8]. The result was a measurable, marked improvement in thermal conductivity.

In the period from roughly 2018 to the present, the field of thermally conductive fluids (TCFs) has rapidly branched out into a multi-faceted domain. The addition of 1D and 2D nanomaterials, such as single-walled CNTs (SWCNTs) and graphene, and the emergence of MXene [9, 10] turned TCFs into fluids that can do many things and really catapulted the field into the forefront of research in three main areas: thermal management and energy storage, electrochemistry, and tribology [10, 11]. This evolution indicates that this field has developed past the proof-of-concept stage, rising into application-specific optimization. Supercapacitors and redox flow batteries are two examples of next-generation energy storage devices made possible by the recent application of INCs as novel electrolytes in electrochemistry [12, 13]. Tribology is also being investigated as a means of employing INCs as environmentally benign, high-performance lubricants to reduce wear and friction [2, 14]. Current research emphasizes sustainability, commercial viability, and cost-effectiveness.

The evolution of key research themes in INCs can be classified into the following three phases:

Phase I: Synthesis, Stability, and Fundamental Characterization *(Early 2000s to Early 2010s)*

Phase II: Mechanistic Insights and Performance Assessment *(Mid-2010s to Late 2010s)*

Phase III: Multifunctionality and Advanced Applications *(Late 2010s to Present)*

The first phase of research was hampered by the difficulty of stabilizing the tiny particles within the highly viscous and electrically charged environment of the IL, when studying the synthesis of nanoparticles in ILs. Coming from this, much of the initial research was devoted to refining the synthesis process, specifically the "two-step" method, which is the most frequently employed method of discharging pre-made nanoparticles into the IL. At the same time, scientists also investigated the basic stabilization mechanism in INCs. According to Nieto de Castro et al., the intrinsic charge and strong Coulombic forces in ILs were found to be vital in preventing nanoparticle agglomeration in INCs, in contrast to aqueous systems in which surfactants are often necessary. The core thermophysical properties, such as thermal conductivity, viscosity, density, and specific heat capacity, were also studied methodically.

During the second phase, the research focus was to figure out the "how" and "why" of the observed property increases and to establish reproducible synthesis procedures and performance assessment. It also became clearer that thermal transfer in these systems could not be effectively understood by classical methods. A more complex image surfaced, and a number of hypothesized mechanisms took the stage. The main topic of debate that began was the exact process that causes the unusual increase in thermal conductivity. Though several theories have been proposed, a universal agreement remains vague. The first theory of *nanoparticle–liquid interfacial layer* is widely accepted at present, in which each nanoparticle is surrounded by a highly organized "solvation layer" of ILs. Based on this theory, a solidified, highly ordered layer of ILs is essentially glued to the nanoparticle surface in this layer and plays the role of a heat transfer facilitator between the particle and the surrounding liquid. The extraordinary capabilities of high-aspect-ratio materials like CNTs and graphene in providing record-breaking enhancements, to some extent, validate the idea that amplifying the surface area of this interfacial layer is the key. The debate still exists on the fact that the exact thickness, structure, and thermal conductivity of this layer are extremely difficult to determine experimentally. The failure of classical theoretical models such as those of Maxwell and Hamilton to accurately predict INC properties stresses this mechanistic gap. Although advanced theoretical frameworks, such as the Murshed, Leong, and Yang (MLY) model, attempt to incorporate interfacial effects, they call for inputs that are experimentally hard or unfeasible to determine, such as the nanolayer thickness (h) and particle clustering factor (ω) [15]. Therefore, this necessitates moving toward highly localized characterization and computational modeling, chiefly molecular dynamics simulations, to convert the theoretical models into accurately prognostic tools.

Another theory is the *nanoparticle micro-convection*. This theory proposes that the Brownian motion of nanoparticles creates micro-convection which enhances the energy transport. This random motion of nanoparticles was expected to induce localized convection at the microscale, creating additional pathways for heat transport. However, many researchers also point out that the high viscosity of ILs severely dampens this Brownian motion, making its role insignificant. Besides these, *nanoparticle clustering mechanism*, i.e., the formation of controlled nanoparticle aggregates, was also suggested as a cause for high-conductivity pathways through the fluid. Though large-scale agglomeration leads to sedimentation and performance degradation, some researchers have proposed that the formation of these wobbly, percolating nanoparticle clusters could produce proficient pathways for phonon transport. This view is however controversial, as it diverges from the goal of achieving long-term colloidal stability. Investigations during this period also included convective heat transfer performance of INCs, their usefulness as direct absorbers in solar collectors, and their performance in heat pipes.

The third phase of research is the investigations on multifunctionality and advanced applications of INCs. This era is currently focusing on application-driven, target-specific technological interventions. Key areas include the following:

- *Advanced Lubrication*: Using INCs as high-performance lubricants in which the IL forms a pliant tribofilm and the nanoparticles play the role of nano-ball bearings or polishing agents that considerably reduce friction and wear.
- *Thermal Energy Storage*: Using INCs as a dual-purpose media for heat storage as well as heat transfer, chiefly in CSP applications.
- *Electrochemical Systems*: Using INCs as advanced electrolytes in batteries and supercapacitors.
- *CO_2 Capture*: Using hybrid INCs, having functionalized nanoparticles (e.g., porous carbons and MOFs) dispersed in task-specific ILs.

The scheme in Figure 1.1 shows the different phases of research development in INC that were discussed above.

Over the past few decades, several pioneering investigations ranging from high thermal conductivity to applications in tribology and solar energy have established the groundwork of INC research. Nieto de Castro et al. investigated MWCNTs dispersed in 1-butyl-3-methylimidazolium bis(trifluoromethylsulfonyl)imide ([C_4mim][NTf_2]) IL [16]. Even with 1 wt.% nanoparticle loading, the thermal conductivity enhanced nearly 65% relative to the base IL. This remarkable observation was far beyond the guess of classical effective medium theories (EMTs) such as the Maxwell model. This presented the first convincing proof that non-classical interfacial phenomena (e.g., nanolayer structuring, micro-convection) may preside over thermal transport in INCs. They also provided baseline data for conductivity and heat capacity for various temperature ranges, providing the empirical foundation for later models. Foundational work by Wang et al. is also pivotal in which surfactant-free synthesis of graphene-based INCs with 18% thermal conductivity increase and high colloidal stability was reported [7]. Later, the effects of nanostructure geometry on conductivity enhancement were quantitatively validated by Jóźwiak et al. [17], and Dzida et al. [18] provided quantitative validation of nanostructure geometry effects on conductivity enhancement, reporting up to 67.8% enhancement for carbon nanostructures.

Recent advances bring in amino acid-based ionic liquids (AAILs) such as 1-butyl-3-methylimidazolium glycinate and arginate along with MWCNTs. These systems demonstrate remarkably low viscosities, high specific heat capacities, and 21–40%

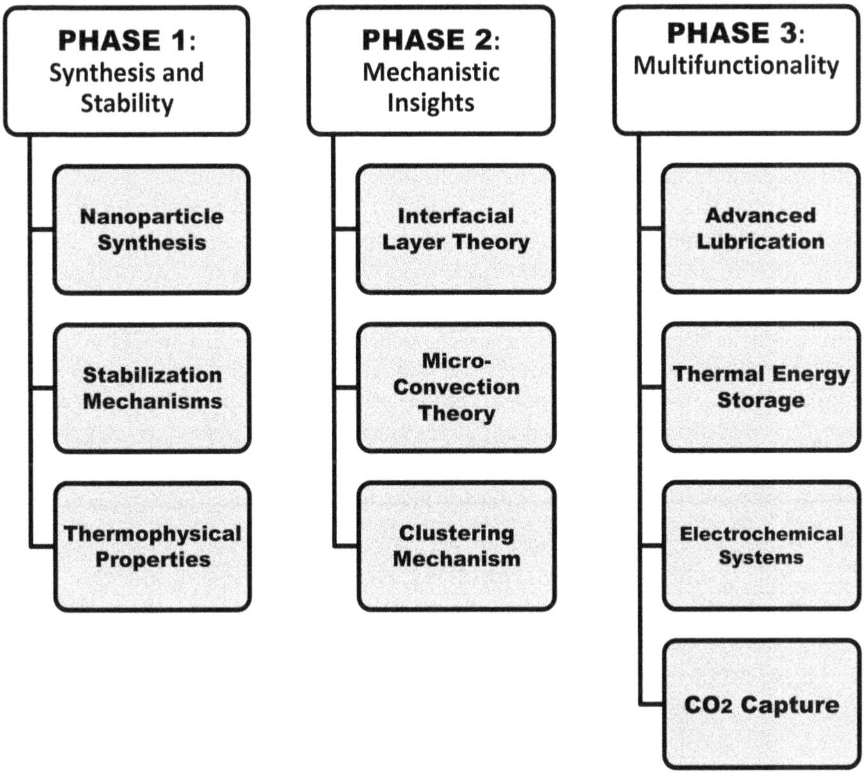

FIGURE 1.1 The scheme shows the different phases of research development in the field of INC.

enhanced thermal conductivity when compared to conventional imidazolium IL–based INCs [19]. The colloidal stability of these INCs exceeded 30 days. This is because the zwitterionic nature of the AAIL anions enhanced hydrogen bonding and electrostatic interactions at the nanoparticle interface. Another breakthrough was the report of sub-zipped CNT INCs in 2022 by Dzida et al. [18]. This study highlights the importance of nanoscale architecture and interface chemistry as design parameters for next-generation heat transfer fluids. In these INCs, three-dimensional percolated thermal bridges through supramolecular connectivity and also nanolayer coalescence were observed within IL matrices. The CNT wall–IL cation covalent interactions and the anion conformational enrichment at tube junctions created enhanced stability and conductivity.

New INCs based on 1-ethyl-3-methylimidazolium acetate and graphene oxide (GO) or graphene nanoplatelets (GNPs) with superior thermal efficiency and colloidal stability (~2 years) that find suitable applications in solar thermal collectors and hybrid photovoltaic-thermal (PV-T) systems were also reported [20]. When the IL was used only as the base fluid instead of water, the results showed a 37.4% increase in efficiency and a 2.5-fold increase in temperature inside the collector. These researches establish INCs as essential working fluids with environmental robustness and high photothermal conversion efficiency for renewable energy systems. It has been discovered that binary ILs enhance the exfoliation and dispersion of metal oxides and MOFs, including TiO_2, ZIF-8@ZIF-67, and other frameworks [21, 22]. The resulting homogeneous nanostructured networks increase catalytic activity and mass transfer rates, expanding the use of INC beyond heat management to include gas treatment and desulfurization catalysis. Several comprehensive reviews have integrated many such experimental insights, which brought to light the influential roles of IL functional groups, nanoparticle morphology, and volume fraction on INC rheology and heat transfer [1, 23–26]. Many important investigations have been documented in several landmark review articles [1]. In 2021, Minea et al. created early stability evaluation guidelines by compiling extensive data on INC synthesis and thermophysical optimization [27]. In order to provide INC criteria for the incorporation of renewable energy, Das et al. [11] and Paul et al. [26] expanded the conversation to solar thermal systems. An important summary of thermal and electrochemical transport in various INC systems was also provided by Faizan et al. [28].

1.2 SYNTHESIS, CHARACTERIZATION, AND STABILIZATION

1.2.1 Preparation Techniques and Rheological Behavior

There are different synthesis methods commonly adopted for INC production, as can be seen in Figure 1.2. The traditional two-step top-down approach for synthesizing INCs involves first synthesizing the nanoparticles and then dispersing them into an IL. This, however, poses the problem of non-uniform dispersion of particles. Uniform dispersion of nanoparticles is practically impossible to achieve without high-energy techniques. Ultrasound is one of the methods that works well against

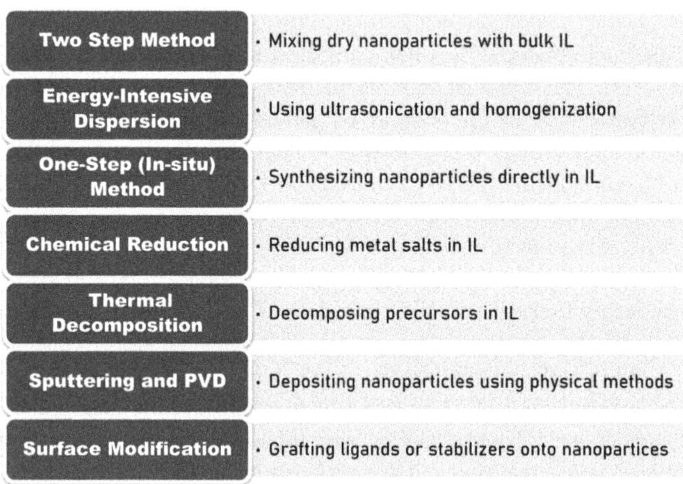

FIGURE 1.2 The scheme shows the different methods commonly adopted for INC production.

the tendency of these particles to agglomerate [29]. Single-step bottom-up synthesis methods, in which nanoparticles are formed within the IL *in situ*, are less common, mainly because of the difficulty of controlling nanoparticle nucleation and growth within the IL medium.

One of the critical challenges in engineering INCs is the impact of rheological properties, particularly viscosity. Adding more nanoparticles to the medium to enhance the thermal performance of these nanofluids would cause an increase in the fluid viscosity [30]. This, in turn, requires higher pumping power for circulation, reducing the overall net heat transfer efficiency of cooling systems [31]. Besides, though the base ILs often display classic Newtonian behavior, the addition of nanoparticles to it makes the resulting INC non-Newtonian in nature [32]. Studies reveal that the range of temperatures over which an INC behaves like a Newtonian fluid is much smaller when compared to that of the pure IL. For example, when loaded with small MWCNTs (≥0.5 wt.%), the INC becomes thixotropic and very hard in nature [33]. This non-Newtonian nature of the INCs presents significant challenges for computational fluid dynamics modeling in industrial heat exchangers.

1.2.2 Long-Term Stability Assessment

A major challenge that researchers face while working with nanofluids is their long-term colloidal stability. The way nanoparticles disperse within the base medium determines their thermophysical properties, rheology, and reliability. Nanoparticle agglomeration and consequent sedimentation often hinder their performance consistency [29]. This is due to the high inherent nanoparticle surface energy which limits the operational lifetime. Consistent performance over time is especially important for high-stakes applications like solar thermal collectors [11].

However, INCs demonstrate significantly improved long-term stability compared to regular nanofluids, which use water or organic oils [4]. The positive and negative ions in the IL base fluid form an ionic shell around them that acts as a protective barrier and prevents them from sticking together. This also reduces the dependence on external chemical dispersants or surfactants, avoiding chemical impurities that might alter the base fluid's intrinsic properties. This in turn makes the synthesis process very much simpler and cheaper.

1.2.3 Rheological Behavior and Viscosity Trends

INCs have both Newtonian and non-Newtonian rheology [34]. With larger nanoparticle loadings or agglomeration levels, shear-thinning (pseudoplastic) behavior appears [23]. Viscosity increases substantially for the majority of stable INCs when nanoparticles are added. However, sheet-like or elongated nanostructures produce lesser increases than spherical particles [27]. Most importantly, rising temperatures cause a significant decrease in viscosity, which improves its pumpability and heat transfer efficiency [35]. Higher temperatures lead to linear shear stress–shear rate relationships and a decrease in the relative viscosity enhancement caused by nanoparticles [27]. All systems show a consistent decrease in viscosity with heating; however, larger particles or higher loadings increase viscosity at a given temperature [36].

These results demonstrate the possibility of creating temperature-responsive fluids that are tailored for dynamic thermal systems. Across various temperatures, linear shear stress increases with shear. However, at higher temperatures, the overall shear stress, and thus the viscosity, is lower, which enables better flow [37].

1.2.4 Characterization Techniques for Ionanocolloids

It is crucial to characterize INCs in order to comprehend the structure–property relationships and maximize performance in mechanical, electrochemical, and thermal applications. Conventional nanofluid characterization procedures need to be modified because of the strong ionic nature and complex interfacial phenomena in ILs. The primary experimental methods for evaluating INC morphology, structure, and thermophysical characteristics are outlined in this section, along with typical problems and best practices to guarantee accuracy and reproducibility.

1.2.4.1 Morphological and Structural Analysis

The micro- and nanoscale morphology of INCs influences their macroscopic parameters. Thus, accurate characterization of particle size, distribution, and interfacial structuring is important. There are numerous techniques used for this purpose, some of which are described in the following:

a. *Microscopy Techniques*: Microscopic techniques, such as transmission electron microscopy (TEM) and scanning electron microscopy (SEM), allow a direct visualization of particle size, shape, and dispersion state within the IL matrices. Atomic-scale resolution of crystalline domains is possible using TEM imaging, whereas topographical insights for dried IL–nanoparticle films can be obtained using SEM. Another well-known technique, atomic force microscopy (AFM), makes nanoscale surface characterization possible. This helps to reveal the nanolayered ion structures and surface roughness.
b. *X-ray Diffraction (XRD)*: Crystallinity and phase composition of nanoparticles within ILs can be determined by XRD analysis. In certain cases, slight shifts in diffraction peaks or variations in crystallite size indicate ion intercalation or IL-induced structural distortion. Furthermore, wide-angle XRD can be used for probing short-range ordering of ions in nano-confined IL environments.
c. *Vibrational Spectroscopy (FTIR and Raman)*: Vibrational spectroscopy is also a very important tool for characterizing INCs. Fourier-transform infrared (FTIR) and Raman spectroscopy provide chemical fingerprints of IL–nanoparticle interactions within the INCs. The presence of a characteristic peak shift or intensity variations, probably in the IL's cation–anion stretching or bending vibrational modes, indicates coordination or adsorption of IL ions on nanoparticle surfaces. For example, a blue shift in vibrational modes of imidazolium ring can disclose the occurrence of any charge–transfer interactions.
d. *Small-Angle X-ray and Neutron Scattering (SAXS/SANS)*: For analyzing inquisitive information of interfacial nanolayer structures and particle–particle correlations within INCs, SAXS and SANS are very helpful. These

techniques can resolve hierarchical structures at scales of 1–100 nm and provide an insight into ion layering, solvation shell thickness, and nanoparticle aggregation behavior inside the INCs under varying temperatures and concentrations.

1.2.4.2 Measurements of Thermophysical Properties

Thermophysical characteristics, such as thermal conductivity, viscosity, density, and heat capacity, determine the ultimate functional performance of INCs in heat transfer or energy storage systems. As can be seen from Figure 1.3, there are many sophisticated techniques currently available for this.

a. *Thermal Conductivity*: The presence of nanoparticles in INCs often augments the heat transport property through percolated thermal pathways or via modifying the IL's phonon and charge-transport mechanisms. For measuring thermal conductivity, methods, such as the transient hot-wire (THW) method, the three-omega (3ω) technique, or laser flash analysis, can be used.

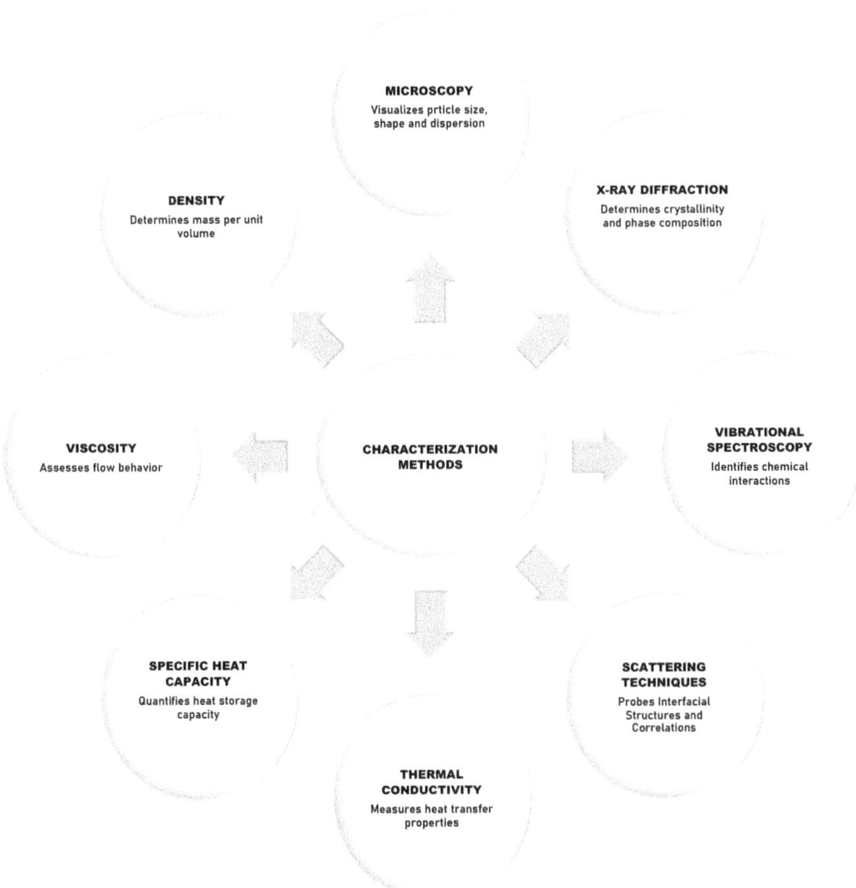

FIGURE 1.3 Schematic summary of major characterization methods for INCs.

The THW method minimizes convection, thereby giving high precision for liquids. For measurement in small sample quantity and at high temperatures, 3ω methods are recommended. For measuring apparent conductivity and diffusivity in solid-like or gelled INCs, laser flash techniques can be used.

b. *Specific Heat Capacity (C_p)*: The heat capacity and thermal stability of INCs can be quantified using differential scanning calorimetry (DSC). DSC thermograms are very helpful to recognize phase transitions, melting point changes, and ionic rearrangements caused by nanoparticle addition. Although the contribution of nanoparticles to C_p is expected to be minimum, they can change the heat storage ability of the INC by changing the IL structuring.
c. *Viscosity (μ)*: Viscosity is always an important design parameter which controls pumping power, flow behavior, and convective heat transfer. Rotational rheometers (cone-and-plate or parallel-plate configurations) can be used to measure viscosity. ILs usually display a non-Newtonian shear-thinning behavior when nanoparticle concentrations surpass the percolation threshold. Temperature-dependent viscosity data are also analyzed to elucidate interfacial bonding strength and intermolecular network formation.
d. *Density (ρ)*: For obtaining accurate, temperature-resolved density measurements of INCs, vibrating-tube densitometers can be used. The density data are crucial for calculating thermodynamic quantities (e.g., enthalpy and specific volume) and also for confirming the most suitable theoretical models for explaining INC behavior. The deviation of density from additive rules can point toward an ion–particle clustering or partial solvation mechanism.

1.2.4.3 Characterization Techniques for Stability

Numerous analytical techniques are available to evaluate INC stability over time and under varying conditions. A few key techniques are discussed here. An inclusive analysis of INC stability requires the integration of several of these analytical techniques to explore the various aspects of dispersion behavior. These techniques collectively offer a multidimensional view of INC stability, correlating microscopic interactions with macroscopic performance. The most widely employed methods are summarized below:

a. *Dynamic Light Scattering (DLS)*: DLS is one of the most efficient tools for monitoring the hydrodynamic diameter of dispersed nanoparticles. This technique helps to observe aggregation phenomena in real time. By analyzing scattered light fluctuations caused by Brownian motion, DLS presents a quantitative data on particle size distribution as well as the temporal evolution of dispersion stability.
b. *Zeta Potential Measurements*: Zeta potential measurements offer insight into the electrostatic stability and surface charge characteristics of nanoparticles dispersed within the IL medium. Though it gives insight into surface charge behavior, interpretation in ILs is complicated due to the high ionic conductivity and low Debye length of ILs, since these factors significantly influence the electrical double layer. Regardless of this complexity, zeta

potential is still an important comparative tool for evaluating trends in surface modification and dispersion behavior.

c. *Visual Observation*: A simple yet effective method is the visual inspection using a time-lapse photography that can qualitatively assess the sedimentation process and any phase separation over extended periods. Even though not a quantitative technique, this approach is also a rapid means to test screen stability and identify systems that demonstrate macroscopic settling or phase stratification.
d. *Centrifugation Tests*: Centrifugation will speed up the sedimentation processes. The application of high centrifugal force can be used to evaluate the sedimentation resistance of INCs. This method enables rapid evaluation of an INC's resistance to phase separation. This test helps to determine short-term stability and also foresee the long-term behavior of INCs under gravitational settings.
e. *Rheological Analysis*: Rheological measurements are widely used to study the flow and deformation behavior of INCs. It measures yield stress, viscosity, and viscoelastic moduli to disclose any transition to gelled or percolated states. Thus, it also helps to identify network formation or particle–particle interactions. It is an oblique but sensitive probe of microstructural evolution within INCs.

By combining time-resolved and structural analyses, researchers can better elucidate the mechanisms that govern the dispersion, aggregation, and long-term reliability of INCs.

1.2.4.4 Challenges and Best Practices

Even though there are numerous characterization tools available for INCs, accurate understanding of the measurement data might need an awareness of several key challenges, as mentioned below:

- *Water and Moisture Sensitivity*: Since most of the ILs are hygroscopic in nature, even trace amounts of water can significantly alter their viscosity, conductivity, and interfacial structures. Therefore, moisture control and vacuum drying conditions must be established well in advance before taking measurements.
- *Temperature and Pressure Dependence*: Temperature and pressure effects have been observed to have nonlinear effects on thermophysical characteristics. Therefore, data reproducibility will be possible only if measurement conditions and thermal history are reported accurately.
- *Measurement Standardization*: Discrepancies in reported data often result from inconsistent protocols or non-standardized calibration steps. Hence, inter-laboratory standards must be established at once for IL and INC testing.

Therefore, after considering all these facts, addressing environmental sensitivity and methodological variability seems to be pivotal for generating reproducible, high-quality data that is essential to progress this field.

1.2.5 Colloidal Stabilization Mechanisms in Ionic Liquids

One of the most significant difficulties in the technological use of INCs is still their stability. The DLVO (Derjaguin–Landau–Verwey–Overbeek) framework effectively predicts colloidal behavior in ordinary nanofluids, but INCs function in a high ionic–strength domain that significantly modifies interparticle interactions. Colloidal stability is defined as a balance between repulsive electrostatic interactions and attractive van der Waals forces in traditional DLVO theory. However, the Debye length is quite short (usually less than 1 Å) in ILs, where ion concentrations can surpass 10 mol L^{-1}, and this effectively suppresses long-range electrostatic repulsion [38]. As a result of DLVO theory's inability to precisely forecast nanoparticle behavior in ILs, additional stabilization models are required.

Electrostatic stabilization caused by surface charge and electrical double-layer repulsion works well for aqueous systems. However, in the case of ILs, the scenario is more complicated. Even if zeta potential values point toward the surface potentials, these readings at times have a contribution from ionic conductivity and electrode polarization also. Therefore, the concept of "zeta potential" itself turns out to be vague in IL-based systems. While some degree of electrostatic contribution cannot be entirely ruled out, it is not considered the dominant stabilization mechanism in IL-based systems [39]. In ILs, oscillatory ion layering close to the surface takes the place of the typical double layer, making the term "zeta potential" itself confusing. Consequently, electrostatic contribution is not regarded as the primary stabilizing mechanism in IL-based systems, even if it cannot be completely ruled out. Instead, steric and solvation stabilization are widely considered the governing mechanisms. These effects occur from structured interfacial nanolayers around the nanoparticle surface which might be of several ions thick. This layer offers a physical obstruction to nanoparticle agglomeration, enabling long-term dispersion possible [38]. Inside a polymer- or surfactant-rich system depletion stabilization may occur. In this case, there exist osmotic pressure differences between polymer-excess and nanoparticle-excess domains [40]. Entropic forces cause the separation of nanoparticles, effectively preventing any agglomeration process.

INCs are however prone to agglomeration, sedimentation, and even gelation even with interfacial stabilization processes, especially in non-equilibrium situations. An important factor is the working temperature of the INC system. Higher temperatures encourage Brownian motion, which can counteract weak attractive forces, but they may also change the interparticle potential landscapes and IL viscosity, which could ultimately result in destabilization [41]. Colloidal behavior is greatly influenced by the concentration of nanoparticles. Stable dispersions prevail at low nanoparticle concentrations. Interparticle networks may form at greater nanoparticle loadings, changing the system's state from liquid-like to gel-like or "ionogel" form [42]. It requires a clear understanding of this shift for optimization of rheological performance and application-specific characteristics of the INCs. Long-term stability however remains a major challenge. While some INCs maintain their stability for several months, others display sudden aggregation within hours, depending on particle surface chemistry, IL composition, and other experimental factors.

1.3 MULTIFUNCTIONAL APPLICATIONS OF INCs

INCs have become a diverse class of multifunctional materials beyond their well-established usage in heat management. Many applications in electrochemistry, tribology, sensing, and gas separations have become possible due to their high ionic conductivity, variable polarity, and great thermal stability. The synergistic effects between ILs and nanoparticles that allow showcasing performance enhancement are not possible with either component alone.

1.3.1 Electrochemical Applications

INCs are used as electrolytes in electrochemical systems. They offer a unique dual functionality. Because ILs are a highly conductive medium, the embedded nanoparticles furnish additional charge-transport pathways and structural stability. Compared with neat ILs, INCs often exhibit enhanced ionic conductivity, improved ion mobility, and better interfacial charge–transfer owing to the high surface area and polarizable interfaces introduced by the nanoparticles [38]. INCs increase the effective electrode–electrolyte interface area in electrical double-layer capacitors (EDLCs), improving capacitance and energy density. Improved ion adsorption and charge–transfer kinetics are two benefits of nanostructured carbons (graphene, CNTs, etc.) distributed in ILs [43]. In addition, ILs allow for higher operating voltages than aqueous systems because they stabilize the electrical double layer across large potential windows.

The non-flammability, broad electrochemical window, and thermal robustness of IL-based electrolytes make them attractive candidates in Li-ion batteries [44]. When incorporated with nanoparticles, the resultant INCs show enhanced cycle stability, reduced dendrite formation, and improved lithium-ion diffusion [45]. By immobilizing anions and raising cation transference numbers, functional fillers like oxide nanoparticles of Al, Si, or Ti can lower polarization losses [46]. INCs are also efficient electrocatalysis media, in which the IL phase assists in ionic conduction and product separation, whereas nanoparticles form active catalytic centers [13, 47]. The unique solvation environment offered by ILs can stabilize intermediates, promote electron transfer, and modify reaction selectivity [48]. Examples of reactions in which the INC's dual catalytic and conductive properties improve overall efficiency are CO_2 reduction, hydrogen evolution, and oxygen reduction reactions [49].

1.3.2 Tribological Applications

Assessing the effectiveness of ILs as lubricants, the ability of these liquids to form a protective tribofilm at the friction interface is one of the key characteristics that establish them as such. The emerging class of INCs now seeks to enhance these tribological properties with their finely dispersed nanoparticles. Due to their strong surface affinity, non-volatility, and thermal stability, INCs serve as high-performance lubricants and anti-wear additives in the field of tribology [50]. The nanoparticles in INC will provide mechanical protection by reducing wear and friction during sliding contact through a "rolling bearing" or mending effect. Meanwhile, the IL component creates an adsorbed boundary film [51]. Nanoparticles have the ability to deform or exfoliate under tribological stress, producing protective tribofilms that reduce adhesive wear. Furthermore, the anti-corrosion behavior is additionally improved by the strong interactions between the polar IL cations and anions and metallic surfaces. Because of these synergistic effects, INCs are promising options for use in extreme lubricating conditions, such as aerospace or microelectromechanical (MEMS) systems.

Experimental investigations show that metal oxide nanoparticles can improve ILs' ability to reduce friction. For example, the addition of copper oxide (CuO) and zinc oxide (ZnO) nanoparticles to the IL [N1888] decreased the friction coefficient, with improvements ranging from 2.4 to 3.5%, depending on concentration [2]. Importantly, though, the anti-wear performance is determined by the particular chemical and mechanical interaction (the lubrication mechanism) of the type of nanoparticle, leading to a wide range of results. Therefore, the manner in which a nanoparticle works is crucial to its anti-wear properties. Zinc oxide (ZnO), for example, significantly increased the anti-wear characteristics of the IL here, around 0.2% reduced the wear scar diameter by 32% compared to the pure IL. It appears that zinc oxide works as a third body that uses pure rolling action to get rid of wear and gives the fluid a boost in anti-wear properties. On the other hand, copper oxide (CuO) added to [N1888] decreased the anti-wear characteristic and in some cases increased the wear width. This is due to a tribosintering effect [52]. This highlights the importance to know exactly how the chosen nanoparticles will behave in the mix, because, the performance of the base IL will decline due to a poor selection of nanoparticles.

1.3.3 Optical, Magnetic, and Sensing Applications

Plasmonic INCs with strong optical tunability are created when noble metal nanoparticles (Au, Ag) are dispersed in ILs [53]. In addition to stabilizing the nanoparticles, the IL medium offers a non-volatile, thermally stable environment for chemical sensing and surface-enhanced Raman scattering (SERS) applications. Plasmon resonance frequencies can be shifted by ILs' high refractive index and adjustable polarity, allowing for customized optical responses [54]. The magnetic INCs (magneto-ionics) that are produced when magnetic nanoparticles like Fe_3O_4 are distributed in ILs show field-responsive behavior, allowing for precise control of motion, alignment, and aggregation [55]. Because of the IL's biocompatibility and adjustable viscosity, these fluids have found use in targeted drug delivery, microactuators, and sealing systems. Luminescent INCs with stable and adjustable emission spectra are produced by embedding quantum dots (QDs) in ILs [56]. By acting as a passivating environment and a dispersion medium, the IL lowers aggregation and photobleaching. These INCs are being investigated for use in bioimaging, photonic sensors, and optoelectronic devices.

1.3.4 Separations and Gas Capture

The blend of selective absorption capacity of ILs and high surface area of nanoparticles has led to the evolution of hybrid INC sorbents for gas separations. Porous materials, such as MOFs or carbon nanostructures, are embedded within ILs to enhance both gas solubility and adsorption kinetics and to overcome mass transfer limitations of pure ILs [57]. INCs reveal outstanding ability for CO_2 capture, showcasing enhanced uptake capacity and faster absorption rates in contrast to neat ILs [58]. In these materials, nanoparticles are the nucleation sites for gas absorption and facilitator of cooperative interactions that weaken C=O bonds in CO_2 molecules. This synergy makes INCs a promising material for carbon capture, gas storage, and environmental remediation technologies.

1.4 CONCLUSIONS

The development of multifunctional working fluids and next-generation heat transfer has advanced significantly with the introduction of IL-based nanofluids, also known as INCs. In order to solve long-standing issues in energy conversion, storage, and management, INCs combine the superior thermal and electrical properties of nanomaterials with the remarkable physicochemical stability and design flexibility of ILs. Fundamental synthesis and characterization have given way to more intricate investigations of structure–property relationships, interfacial phenomena, and application-specific optimization during the last 20 years.

Even with the significant advancements, a number of scientific and engineering problems still remain unresolved. The main concerns are the scalability of synthesis, non-Newtonian flow behavior, measurement standardization, and long-term colloidal stability. In order to accurately predict thermophysical performance under dynamic conditions, integrated modeling approaches that combine molecular dynamics, sophisticated simulations, and empirical validation are necessary, as theoretical understanding still lags behind experimental findings.

The chapter emphasizes that a multidisciplinary strategy bridging materials science, colloidal chemistry, and thermal engineering is crucial for the future development of INCs. For solar thermal harvesting, electrochemical storage, and tribological applications, strategic developments in nanostructure design, IL functionalization, and hybrid fluid formulation should result in customized systems with exceptional performance. Ultimately, INCs hold the possibility to become groundbreaking materials in the search for energy-efficient and environmentally conscious technologies, marking a transition from laboratory-level innovation to industrially viable reality.

ABBREVIATIONS

AAILs	amino acid–based ionic liquids
AFM	atomic force microscopy
Al_2O_3	aluminum oxide (use consistently as Al_2O_3)
Au	gold (element symbol)
CNT/CNTs	carbon nanotube(s) (also MWCNT and SWCNT used below)
CSP	concentrated solar power
DLS	dynamic light scattering
DSC	differential scanning calorimetry
EDLCs	electrical double-layer capacitors

FTIR	Fourier-transform infrared spectroscopy
GNPs	graphene nanoplatelets
GO	graphene oxide
IL(s)	ionic liquid(s)
INC(s)	ionanocolloid(s)
MLY model	Murshed, Leong, and Yang model (spell out on first use)
MWCNT(s)	multi-walled carbon nanotube(s)
MOF(s)	metal–organic framework(s)
PV-T	photovoltaic-thermal systems
SAXS/SANS	small-angle X-ray scattering/small-angle neutron scattering
SEM	scanning electron microscopy
SERS	surface-enhanced Raman scattering
SWCNT(s)	single-walled carbon nanotube(s)
TEM	transmission electron microscopy
THW	transient hot-wire method (thermal conductivity)
wt.%	weight percent
3ω	three-omega thermal measurement technique

REFERENCES

1. Fabre, E. and S.M.S. Murshed, *A comprehensive review of thermophysical properties and prospects of ionanocolloids in thermal energy applications*. Renewable and Sustainable Energy Reviews, 2021. **151**: p. 111593.
2. Ta, T.-N., S.-Y. Chern, and J.-H. Horng, *Tribological behavior of ionic liquid with nanoparticles*. Materials, 2021. **14**(21): p. 6318.
3. Al-Bodour, A., et al., *Ionic Liquids as Multidimensional Materials: A Review from Fundamentals to Applications*. 2025, Energy & Fuels.
4. Minea, A.A. and S.S. Murshed, *A review on development of ionic liquid based nanofluids and their heat transfer behavior*. Renewable and Sustainable Energy Reviews, 2018. **91**: p. 584–599.
5. Bayda, S., et al., *The history of nanoscience and nanotechnology: From chemical–physical applications to nanomedicine*. Molecules, 2019. **25**(1): p. 112.
6. Shiflett, M.B., J.W. Magee, and D. Tuma, *Important developments in the history of ionic liquids from academic curiosity to commercial processes and products*, in *Commercial Applications of Ionic Liquids*. 2020, Springer. p. 3–29.
7. Wang, F., et al., *Surfactant-free ionic liquid-based nanofluids with remarkable thermal conductivity enhancement at very low loading of graphene*. Nanoscale Research Letters, 2012. **7**(1): p. 314.
8. Wang, B., et al., *Gold-ionic liquid nanofluids with preferably tribological properties and thermal conductivity*. Nanoscale Research Letters, 2011. **6**(1): p. 259.
9. Jóźwiak, B., et al., *Remarkable thermal conductivity enhancement in carbon-based ionanofluids: Effect of nanoparticle morphology*. ACS Applied Materials & Interfaces, 2020. **12**(34): p. 38113–38123.
10. Cavieres, J., et al., *Ionanofluids based on ionic liquid mixtures, a new approach as an alternative material for solar energy storage*. Journal of Molecular Liquids, 2022. **351**: p. 118677.
11. Das, L., et al., *State-of-the-art ionic liquid & ionanofluids incorporated with advanced nanomaterials for solar energy applications*. Journal of Molecular Liquids, 2021. **336**: p. 116563.

12. Joseph, A., et al., *Ionic liquid and ionanofluid-based redox flow batteries—A mini review.* Energies, 2022. **15**(13): p. 4545.
13. Pereira, J., et al., *Nanofluids and ionic fluids as liquid electrodes: An overview on their properties and potential applications.* Processes, 2023. **11**(11): p. 3189.
14. Yang, Z., et al., *Frictional characteristics and tribological mechanisms of ionic liquid lubricants in ceramic tribo-systems.* Materials, 2025. **18**(19): p. 4504. https://doi.org/10.3390/ma18194504
15. Murshed, S.M.S., K.C. Leong, and C. Yang, *A combined model for the effective thermal conductivity of nanofluids.* Applied Thermal Engineering, 2009. **29**(11): p. 2477–2483.
16. Nieto de Castro, C.A., et al., *Thermal properties of ionic liquids and ionanofluids of imidazolium and pyrrolidinium liquids.* Journal of Chemical & Engineering Data, 2010. **55**(2): p. 653–661.
17. Jóźwiak, B., G. Dzido, and E. Zorębski, *Remarkable thermal conductivity enhancement in carbon-based ionanofluids: Effect of nanoparticle morphology.* 2020. **12**(34): p. 38113–38123.
18. Dzida, M. and S. Boncel, *High-performance ionanofluids from subzipped carbon nanotube networks.* 2022. **14**(45): p. 50836–50848.
19. Chandra, A., et al., *Heat transfer fluids: Amino acid anion ionic liquid based ionanofluids with remarkable thermal conductivity and low viscosity†.* RSC Advances, 2025. **15**(29): p. 23146–23164. https://doi.org/10.1039/d5ra03752b
20. Moulefera, I., et al., *Novel application for graphene oxide-based ionanofluids in flat plate solar thermal collectors.* Scientific Reports, 2024. **14**(1): p. 17610.
21. Liu, F., et al., *The enhancement effect of core-shell ZIF-8@ZIF-67 nanoparticles on the desulfurization performance of ionic liquid based nanofluid systems.* Colloids and Surfaces A: Physicochemical and Engineering Aspects, 2025. **726**: p. 137882.
22. Mim, M., et al., *Evaluation of novel binary imidazolium ionic liquid-based WO_3/MgO nanocomposite for light-to-thermal energy conversion and storage – A preliminary study.* Journal of Ionic Liquids, 2025. **5**(2): p. 100172.
23. Jóźwiak, B. and S. Boncel, *Rheology of ionanofluids – A review.* Journal of Molecular Liquids, 2020. **302**: p. 112568.
24. Oster, K., et al., *Ionic liquid-based nanofluids (ionanofluids) for thermal applications: An experimental thermophysical characterization.* Pure and Applied Chemistry, 2019. **91**(8): p. 1309–1340.
25. Lingala, S.S., *Ionic-liquid-based nanofluids and their heat-transfer applications: A comprehensive review.* ChemPhysChem, 2023. **24**(22): p. e202300191.
26. Paul, T.C., et al., *A critical review on the development of ionic liquids-based nanofluids as heat transfer fluids for solar thermal energy.* Processes, 2021. **9**(5): p. 858.
27. Minea, A.A. and S.M. Sohel Murshed, *Ionic liquids-based nanocolloids – A review of progress and prospects in convective heat transfer applications*, 2021. **11**(4): p. 1039. https://doi.org/10.3390/nano11041039
28. Faizan, M., R. Ahmed, and H.M. Ali, *A critical review on thermophysical and electrochemical properties of ionanofluids (nanoparticles dispersed in ionic liquids) and their applications.* Journal of the Taiwan Institute of Chemical Engineers, 2021. **124**: p. 391–423.
29. Jóźwiak, B., et al., *Effect of ultrasonication time on microstructure, thermal conductivity, and viscosity of ionanofluids with originally ultra-long multi-walled carbon nanotubes.* Ultrasonics Sonochemistry, 2021. **77**: p. 105681.
30. Chandra, A., et al., *Heat transfer fluids: Amino acid anion ionic liquid based ionanofluids with remarkable thermal conductivity and low viscosity.* RSC Advances, 2025. **15**(29): p. 23146–23164.

31. Razzaq, I., et al., *Nanofluids for advanced applications: A comprehensive review on preparation methods, properties, and environmental impact.* ACS Omega, 2025. **10**(6): p. 5251–5282.
32. de Castro, C.A.N., et al., *Possible new heat transfer fluid: The ionanofluid of 1-Ethyl-3-methylimidazolium dicyanamide + nano-titanium oxide—Studying its thermal conductivity and viscosity.* Journal of Chemical & Engineering Data, 2024. **69**(6): p. 2227–2235.
33. Dzida, M., et al., *High-performance ionanofluids from subzipped carbon nanotube networks.* ACS Applied Materials & Interfaces, 2022. **14**(45): p. 50836–50848.
34. Hothar, M., B. Sundén, and Z. Wu, *Rheological behaviour of IoNanofluids based on [emim][DCA] and [emim][TCM].* Journal of Molecular Liquids, 2022. **348**: p. 118064.
35. Soman, D.P., et al., *Impact of viscosity of nanofluid and ionic liquid on heat transfer.* Journal of Molecular Liquids, 2019. **291**: p. 111349.
36. Main, K.L., et al., *Nanoparticles size effect on thermophysical properties of ionic liquids based nanofluids.* Journal of Molecular Liquids, 2021. **343**: p. 117609.
37. Bakthavatchalam, B. and K. Habib, *Optimization of thermophysical and rheological properties of MXene ionanofluids for hybrid solar photovoltaic/thermal systems.* 2021. **11**(2): p. 320. https://doi.org/10.3390/nano11020320
38. He, Z. and P. Alexandridis, *Nanoparticles in ionic liquids: interactions and organization.* Physical Chemistry Chemical Physics, 2015. **17**(28): p. 18238–18261.
39. Hayes, R., G.G. Warr, and R. Atkin, *Structure and nanostructure in ionic liquids.* Chemical Reviews, 2015. **115**(13): p. 6357–6426.
40. Mmbuji, A.O., et al., *Nanoparticle-assisted surfactant/polymer formulations for enhanced oil recovery in sandstone reservoirs: From molecular interaction to field application.* Energy & Fuels, 2023. **37**(22): p. 17094–17112.
41. Tadros, T., *Interparticle interactions in concentrated suspensions and their bulk (Rheological) properties.* Advances in Colloid and Interface Science, 2011. **168**(1–2): p. 263–277.
42. Zhao, N., et al., *Liquid crystal self-assembly of halloysite nanotubes in ionic liquids: A novel soft nanocomposite ionogel electrolyte with high anisotropic ionic conductivity and thermal stability.* Nanoscale, 2016. **8**(3): p. 1545–1554.
43. Ye, Y., et al., *Enhanced ion transport in polymer–ionic liquid electrolytes containing ionic liquid-functionalized nanostructured carbon materials.* Carbon, 2015. **86**: p. 86–97.
44. Tu, H., et al., *Solvation and interfacial chemistry in ionic liquid based electrolytes toward rechargeable lithium-metal batteries.* Journal of Materials Chemistry A, 2024. **12**(48): p. 33362–33391.
45. Hu, Z., X. Zhang, and S. Chen, *A graphene oxide and ionic liquid assisted anion-immobilized polymer electrolyte with high ionic conductivity for dendrite-free lithium metal batteries.* Journal of Power Sources, 2020. **477**: p. 228754.
46. Li, M., et al., *Application of the imidazolium ionic liquid based nano-particle decorated gel polymer electrolyte for high safety lithium ion battery.* Electrochimica Acta, 2018. **284**: p. 188–201.
47. Eshetu, G.G., et al., *Ionic liquids as tailored media for the synthesis and processing of energy conversion materials.* Energy & Environmental Science, 2016. **9**(1): p. 49–61.
48. Lee, J.W., et al., *Toward understanding the origin of positive effects of ionic liquids on catalysis: Formation of more reactive catalysts and stabilization of reactive intermediates and transition states in ionic liquids.* Accounts of Chemical Research, 2010. **43**(7): p. 985–994.
49. Chen, K., et al., *Functions and performance of ionic liquids in enhancing electrocatalytic hydrogen evolution reactions: A comprehensive review.* RSC Advances, 2022. **12**(30): p. 19452–19469.

50. Maurya, U., V. Vasu, and D. Kashinath, *Ionic liquid-nanoparticle-based hybrid-nano-lubricant additives for potential enhancement of tribological properties of lubricants and their comparative study with ZDDP.* Tribology Letters, 2022. **70**(1): p. 11.
51. Saurín, N., et al., *Ionic nanofluids in tribology.* Lubricants, 2015. **3**(4): p. 650–663.
52. Roshan, M.S., et al., *Nanomaterial-enhanced greases: A review of tribological and rheological advancement.* Journal of Tribology, 2025. **148**(2).
53. Pereira, J., A. Moita, and A. Moreira, *Noble nanofluids and their hybrids for heat transfer enrichment: A review and future prospects coverage* . Applied Sciences, 2023. **13**(17): p. 9568.
54. Aravind, I., et al., *Voltage-induced modulation of interfacial ionic liquids measured using surface plasmon resonant grating nanostructures.* The Journal of Chemical Physics, 2024. **161**(3).
55. Rodríguez-Arco, L., et al., *Steric repulsion as a way to achieve the required stability for the preparation of ionic liquid-based ferrofluids.* Journal of Colloid and Interface Science, 2011. **357**(1): p. 252–254.
56. Gonçalves, H.M.R., et al., *Non-Newtonian thermosensitive nanofluid based on carbon dots functionalized with ionic liquids.* Small, 2020. **16**(28): p. 1907661.
57. Li, X., et al., *Ionic liquids functionalized MOFs for adsorption.* Chemical Reviews, 2023. **123**(16): p. 10432–10467.
58. Pardo, F., et al., *Integration of stable ionic liquid-based nanofluids into polymer membranes. Part II: Gas separation properties toward fluorinated greenhouse gases.* Nanomaterials, 2021. **11**(3): p. 582.

2 Synthesis, Electrochemical Properties, and Energy Applications of Ionic Liquid-Based Nanofluids

Saleja Sahoo, Pooja Mohapatra, Shreelata Behera, Annada Mishra, Lipsa Shubhadarshinee, Bigyan Ranjan Jali, P. S. Rama Sreekanth, Ajaya Kumar Behera, Bibhu Prasad Sahoo, and Aruna Kumar Barick

2.1 INTRODUCTION

Ionanofluids (INFs) are complex systems consisting of dispersions of nanomaterials within an ionic liquid (IL) carrier medium and are currently projected as a promising approach for sustainable developments [1]. Nieto de Castro *et al.* coined the term "Ionanofluid" in 2009. INFs can be synthesized through different approaches reported in the literature, such as one-step and two-step techniques, as well as polymerization-based methods. These routes provide flexibility in tailoring the nanofluid properties for specific applications. ILs are liquid salts totally composed of ions with melting points below 100°C. These are prepared by incorporating nanomaterials into the ILs that enhance the thermophysical, optical, and energy storage properties of the base ILs. INFs, also referred to as NEILs, are a novel class of nanofluids where the ILs act as the base fluid and nanoparticles serve as the dispersant. Owing to the valuable properties of the ILs compared to the conventional working fluids, their use in research and applications is steadily increasing. In recent decades, extensive studies have been conducted on nanoparticles, leading to the development of various new nanomaterials with excellent thermal properties, including two-dimensional nanomaterials like graphene and MXene along with SWCNTs and MWCNTs, which have emerged as the most promising materials discovered to date [1–4]. IL-based nanofluids (ILNFs) significantly improve their physical properties (density and viscosity) and thermal performance through the dispersion of nanoparticles in ILs. ILNFs are a superior alternative compared to conventional nanofluids because of their enhanced characteristics, which enable their various applications like redox flow batteries (RFBs), dye-sensitized solar

 DOI: 10.1201/9781003595250-2

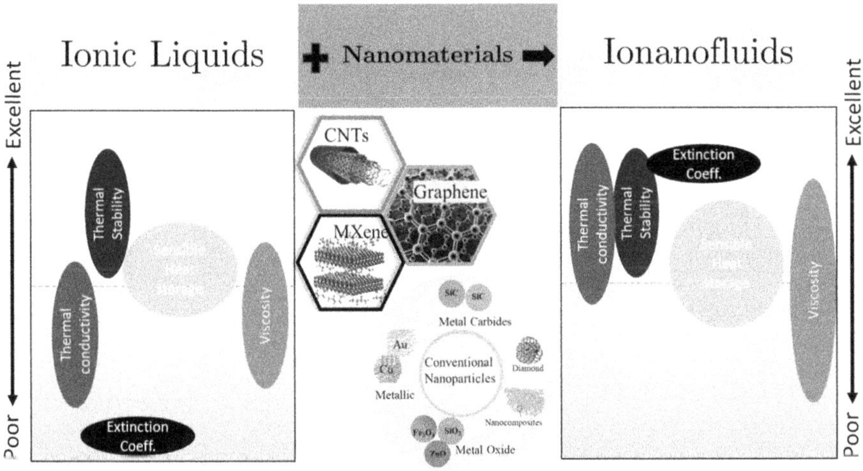

FIGURE 2.1 Schematic diagram of various components of ILNFs [3].

cells (DSSCs), and fuel desulfurization processes (FDP) [5]. At Argonne National Laboratory in 1995, researchers proposed the term nanofluid, which is defined as a synthesized fluid formed by incorporating nanoparticles into a base liquid [6]. The different components of the ILNFs are represented in Figure 2.1.

It has been suggested by researchers that the thermophysical properties of CSP systems can be improved by using nanoparticles in ILs [7–19]. In this context, Chereches *et al.* [7] examined viscosity, specific heat, and electrical conductivity of ILNFs, which were prepared by dispersing Al_2O_3 nanoparticles (0.05–10 wt.%) in IL $[C_2mim][CH_3SO_3]$. Their results showed that the viscosity increased by 250% in ILNFs at a loading of 10 wt.% of nanoparticles. Ferreira *et al.* have seen an enhancement in thermal conductivity, viscosity, and thermal stability of nanofluids formed from MWCNT and phosphonium-containing ILs [8]. Further work has shown the improvement of both density and specific heat in ILNFs formed from imidazolium-type ILs and MWCNTs [9–11]. However, when the MWCNTs are functionalized with carboxylic acid groups, the resulting nanofluids show lower viscosity than the original IL. Alizadeh *et al.* have investigated the viscosity, electrical conductivity, and surface tension of nanofluids formed from ($[Bmim][PF_6]$) IL and graphene [13].

2.2 METHODS OF SYNTHESIS

The stable ILNFs are prepared by dispersing the nanoparticles (MWCNTs and GO) in the ILs, thereby improving their thermophysical properties, including density, viscosity, and thermal conductivity [8, 14, 15]. For example, according to França *et al.*, an increase in the thermal conductivity of 5–35% for the ILNFs containing MWCNTs has been observed depending upon the ILs and concentration of the nanoparticles. The nanoparticles are added to the base ILs using a two-step process, and ILNFs are produced after ultrasonication to provide the uniformity

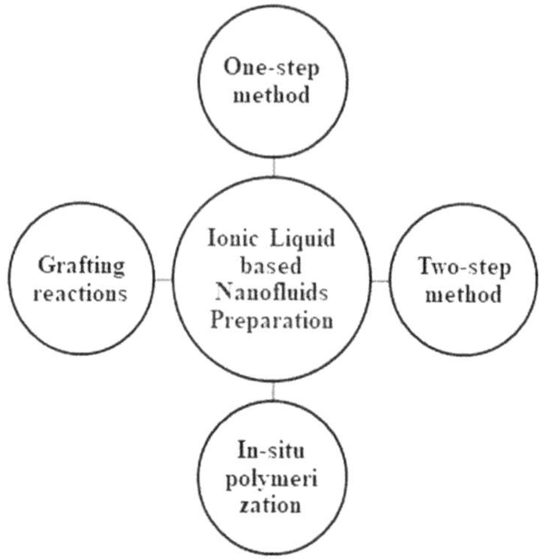

FIGURE 2.2 Schematic diagram of various methods of synthesis of ILNFs.

and stability of the resulting ILNFs [10, 20]. ILNFs can be synthesized through one-step and two-step processes [21, 22], as well as specialized processes such as in situ polymerization and grafting reactions. The above methods ensure the uniform dispersion of the nanoparticles in the base fluid and the improvement in the thermal property. However, the selection of the synthesis method will affect the particle size range of the nanofluids, their overall performance, and their stability. The surface functionalization of the nanoparticles is also generally used to avoid agglomeration and improve compatibility with the ILs. The various methods used for the synthesis of the ILNFs are represented in Figure 2.2.

2.2.1 One-Step Method

In this technique, nanoparticles are synthesized directly within the ILs, promoting uniform dispersion and minimizing agglomerations. The glassware used in the experiment was cleaned with freshly mixed nitric acid and hydrochloric acid (1:3, v/v), rinsed with distilled water, and dried before use. A solution containing TA ($C_{76}H_{52}O_{46}$) (1% w/w) and a stabilizing agent was rapidly added to a gold precursor solution ($HAuCl_4$, 0.01% w/w) under vigorous stirring. The total reaction volume was kept at 50 ml. As the reaction continued, a colour change from yellow to red indicated the formation of AuNFs. The reaction was typically completed within 30 min, and the final product was stored at 4°C [23].

Figure 2.3(a) & (b) shows the effect of different stabilizers on UV–Vis–NIR spectra and the morphologies of AuNFs respectively. Only the sample stabilized with [Aemim][Br] showed a significant absorption peak in the higher wavelength region 800–1100 nm, suggesting the presence of non-spherical nanoparticles. Other stabilizers, i.e., [Cmmim][Cl], NaBr, PVP, and trisodium citrate ($Na_3C_6H_5O_7$), showed no

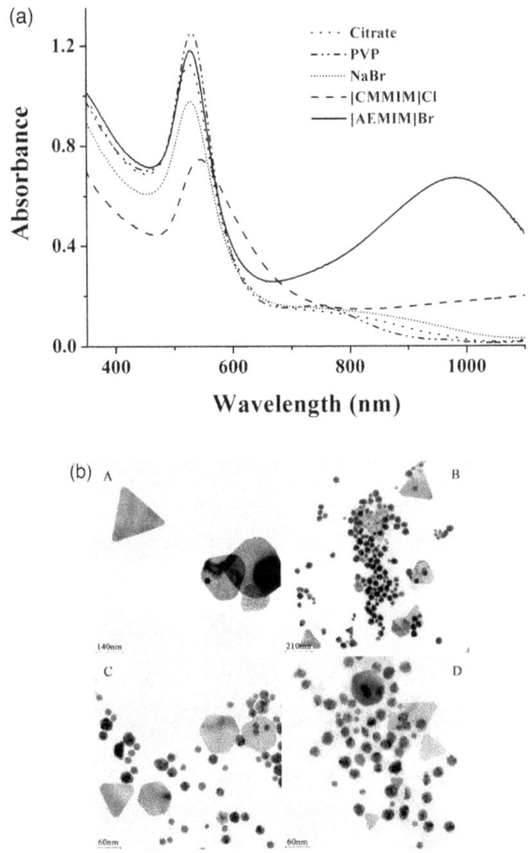

FIGURE 2.3 (a) Effect of different stabilizers on UV–Vis–NIR spectra of AuNFs and (b) The figures A, B, C and D shows the TEM images of synthesized AuNFs [23].

such absorbance, implying primarily spherical or regular-shaped nanoparticles. The absence of absorption at the longer wavelength region in NaBr-stabilized samples confirmed that the bromide ion was not responsible for the non-spherical growth. Instead, the imidazole cation, particularly with amino functionality, was crucial in directing anisotropic growth due to the strong Au–N interactions [24, 25].

Figure 2.3(b) shows micrographs of four samples, illustrating the influence of volume on the particle morphology, in contrast to the stabilizer effects on AuNFs, as shown in Figure 2.3(a). It was observed that the formation of AuNFs sol using 0.10 ml of TA rarely occurred. Figure 2.3(b) shows that varying the amount of reductant alters the morphology of the AuNFs. In Figure 2.3(b(A)), a few triangular structures are scattered across the substrate. As the amount of reductant increased, more sol particles formed and started to aggregate, as shown in Figure 2.3(b(B)), while higher TA volumes further reduced the particle size and introduced polygonal and irregular polyhedral structures [23].

2.2.2 Two-Step Method

In the two-step method, ILNFs are initially synthesized by the dispersion of nanoparticles into the ILs followed by homogenization through ultrasonication, which commonly enhances the dispersion stability [26]. Figure 2.4 depicts the two-step synthesis method of ILNFs and also outlines various techniques used to ensure the stable dispersion of nanoparticles in base ILs. During the preparation of ILNFs, careful handling of the sonication process is of the utmost importance, as higher amplitudes, prolonged exposure, or use of probe-type ultrasonication can degrade the physical and chemical integrity of the materials. This degradation can change the microstructure and surface of the nanoparticles, such as in CNTs. Similarly, long-term sonication may lead to the loss of solvent, resulting in the altered concentration of nanoparticles [20]. The use of surfactants has to be approached with caution since they may be subject to degradation or lose their effectiveness in intermediate to high-temperature conditions. In the final analysis, an evaluation of the stability of ILNFs prepared will be necessary and should be carried out through a comprehensive stability analysis. Key characterization techniques include zeta potential measurement, UV–visible spectroscopy, particle size distribution via DLS, and morphological analysis through TEM/SEM. The stability evaluation methods for ILNFs are similar to those commonly employed for conventional nanofluids [27–29].

For the synthesis of ILNFs through the one-step and two-step methods, each approach has its own advantages and disadvantages. After comparing these two methods, it was concluded that the one-step method minimizes the agglomeration of nanoparticles by avoiding transportation, storage, drying, and dispersion processes. However, it requires low vapour pressure fluids and is difficult to recognize all the species present in solution. Also, this process cannot produce large quantities of nanofluids. On the other hand, the two-step method is a practical and cost-effective process, but it offers little or no stability due to the aggregation of powders caused by strong Van der Waals forces. It also leads to internal stress, surface defects, and contamination and requires the use of surfactants [28].

2.2.3 In Situ Polymerization

This approach involves polymerizing monomers within ILs containing nanoparticles, leading to ILNFs with tailored properties. A novel PIL magnetic nanocomposite was derived based on the Radziszewski reaction using in situ polymerization [30].

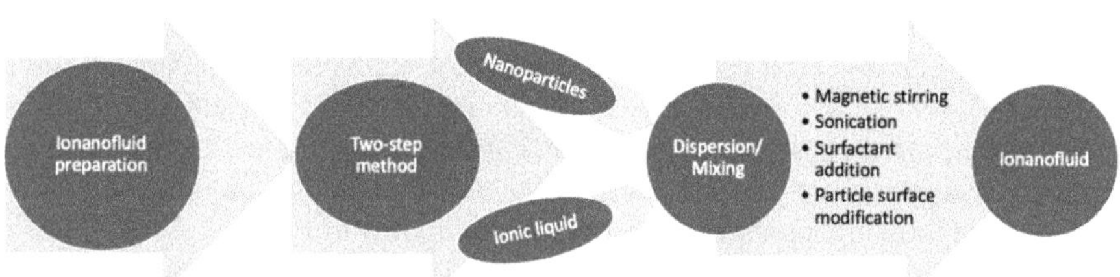

FIGURE 2.4 Schematic representation of synthesis of ILNFs [20].

This method streamlines the synthesis process by minimizing reagent usages and shortening the overall reaction time. As depicted in Figure 2.5(a), the bulk synthesis takes place via the reaction of diamine compounds with glyoxal and formaldehyde under acidic conditions. In this methodology, the employed magnetic nanoparticles (Fe_3O_4) are functionalized with amine groups, which can act as polyamines in the Radziszewski reaction, following specific synthetic routes depicted in Figure 2.5(b) and (c). The synthesized material was comprehensively characterized through a range of analytical methods, and its proper application as a sorbent for NSAIDs, i.e., ketoprofen, naproxen, and salicylic acid in saliva, was explored.

Figure 2.6(a) analyses the IR spectra of Fe_3O_4@NH_2 nanoparticles and nanocomposites. The characteristic absorption bands are observed at 585–632 cm^{-1}, corresponding to the Fe–O–Fe stretching vibrations, at 1000–1200 cm^{-1} for both Si–O and Si–O–H vibrations, at 1625 cm^{-1} for the N–H bending of amino groups, and in the 3400 cm^{-1} region for N–H stretching vibrations, as shown in Figure 2.6(a(a)). These peaks confirm the successful surface modification of the MNPs with APTES. The IR spectrum of the nanocomposites shown in Figure 2.6(a(b)) exhibits two additional absorption peaks at 1382 and 1558 cm^{-1}, which are due to the C–N stretching vibration of the imidazolium ring and the C=O stretching of the acetate group, respectively. These two new bands confirm the successful functionalization of the MNP surface with PILs. Furthermore, the characteristic bands associated with Fe_3O_4@NH_2, such as those for Fe–O bonds (585–632 cm^{-1}) and Si–O/Si–O–H vibrations (1000–1200 cm^{-1}), are still present. Additional weak peaks at 2860 cm^{-1} and 2936 cm^{-1} are attributed to the symmetric and asymmetric stretching vibrations of methylene (–CH_2–) groups, respectively [32].

TEM analysis provided insights into the morphology and size of the nanoparticles and nanocomposites. As illustrated in Figure 2.6(b(a)), the Fe_3O_4@NH_2 nanoparticles exhibit uniformly dispersed spherical-shaped nanostructures with an average diameter of approximately 10 nm. In contrast, Figure 2.6(b(b)) reveals that the nanocomposites exhibit strong agglomeration of nanoparticles, indicating that the PILs act as a binding or entanglement agent.

2.2.4 Grafting Reactions

Chemical grafting of ILs onto nanoparticles can enhance compatibility and dispersion stability within the IL matrix. The IL–GO hybrid nanofillers were synthesized during a simple grafting reaction between imidazole-based ILs and GO [33]. To prepare the IL–GO, a grafting reaction was carried out between amino-functionalized ILs and GO. Specifically, 100 mg of GO was dispersed in water using ultrasonication for 1 h to create a uniform suspension. To enhance the grafting process, an excess amount of IL was introduced. A solution containing 20 mg of ILs was introduced into the GO suspension, and the resulting mixture was stirred continuously and heated at 90°C for 24 h. The IL–GO was then collected via centrifugation, repeatedly washed with deionized water, and then dried in a vacuum oven at 50°C [34, 35].

The FTIR spectra of GO and two GO–ILs, namely GO-[Aemim][Br] and GO-[Apmim][NTf_2], are presented in Figure 2.7(a(a–c)). In Figure 2.7(a(a)), a

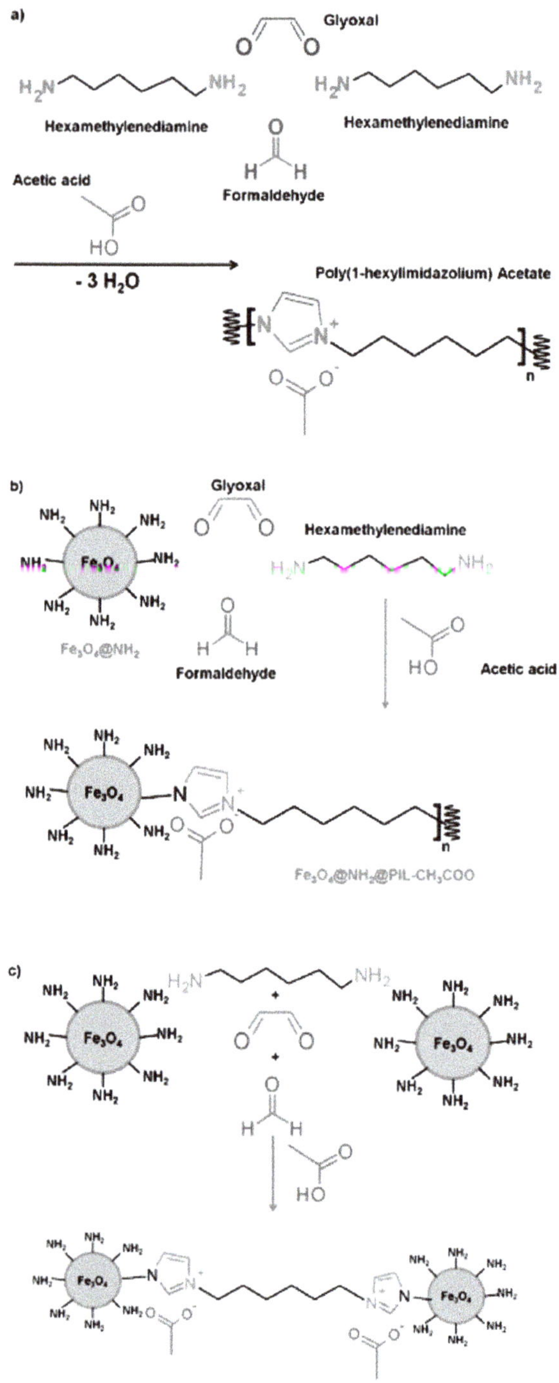

FIGURE 2.5 Schematic diagram of PIL synthesis via the Radziszewski reaction: (a) PILs and (b and c) corresponding nanocomposites [31].

prominent peak around 3500 cm^{-1} assigned to the stretching vibration of –OH groups present on the surface of GO. The peak observed at 1730 cm^{-1} is indicative of the C=O stretching vibration in carboxylic groups, while the one at 1623 cm^{-1} is attributed to the C=C bonds within the aromatic structures [36, 37]. In the spectra of the GO–IL shown in Figure 2.7(a(b) and (c)), the peak at 1167 cm^{-1} is related to asymmetric in-plane stretching vibrations of the imidazolium ring [38]. Additionally, the broad peak near 1618 cm^{-1} and a distinct peak at 1652 cm^{-1} are associated with the amide (–CONH–) functional group [39]. These spectral features confirm the successful grafting of the ILs onto the surface of GO.

(a)

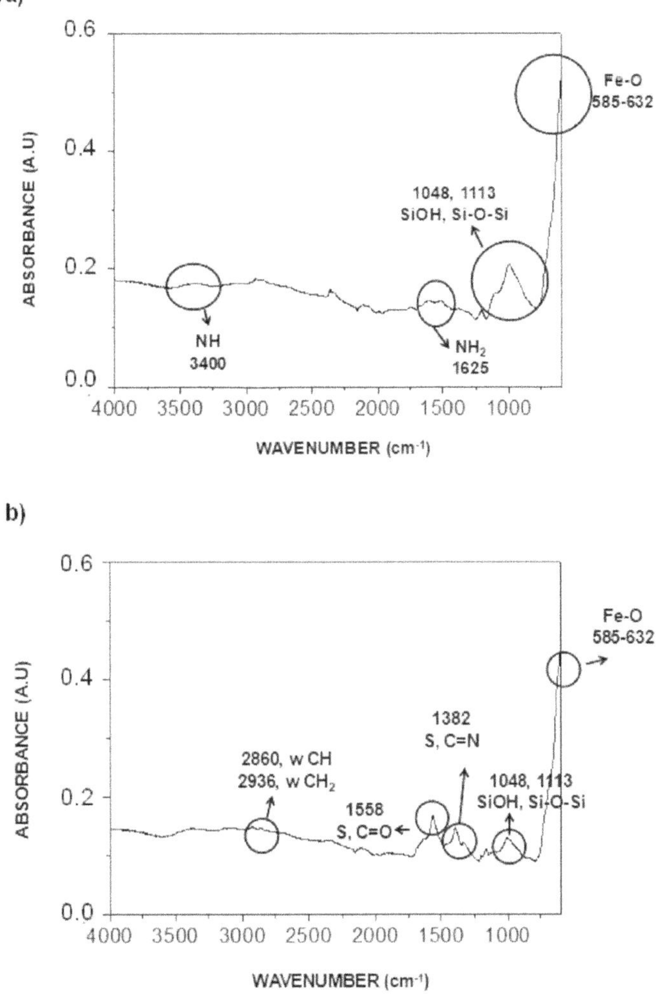

FIGURE 2.6 (a) ATR-FTIR spectra of (a) nanoparticles and (b) nanocomposites, and (b) TEM micrographs of (a) nanoparticles and (b) nanocomposites [31]. *(Continued)*

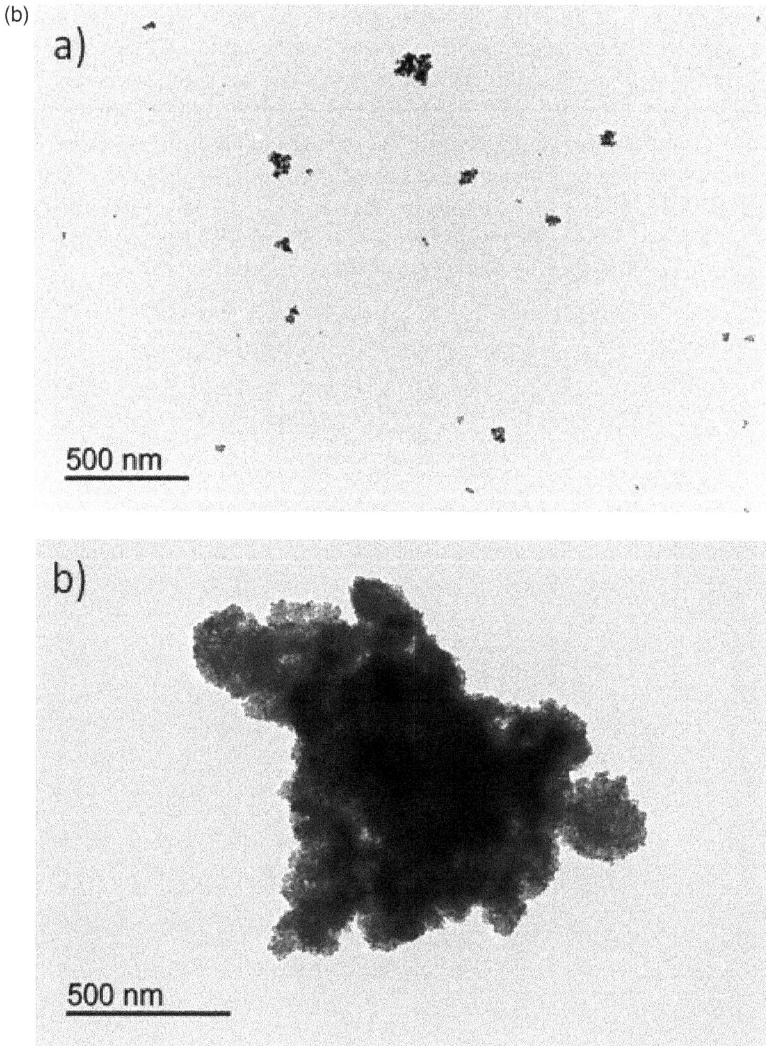

FIGURE 2.6 *(Continued)*

Raman spectroscopy was used to investigate structural changes in GO after IL functionalization, as shown in Figure 2.7(b). GO showed two prominent peaks at 1348 and 1580 cm^{-1} due to D-band and G-band, corresponding to the vibration of sp^3 carbon atoms from the functional groups and the in-plane vibration of sp^2 carbon atoms, respectively [42]. The IL–GO exhibited similar bands with small shifting of peak positions. Especially, the ratio of intensities of the D- to G-bands (I_D/I_G) enhanced from 0.81 for GO to 1.12 for IL–GO [43], revealing an improvement in structural disorders due to the bulky imidazole ring of ILs. This higher

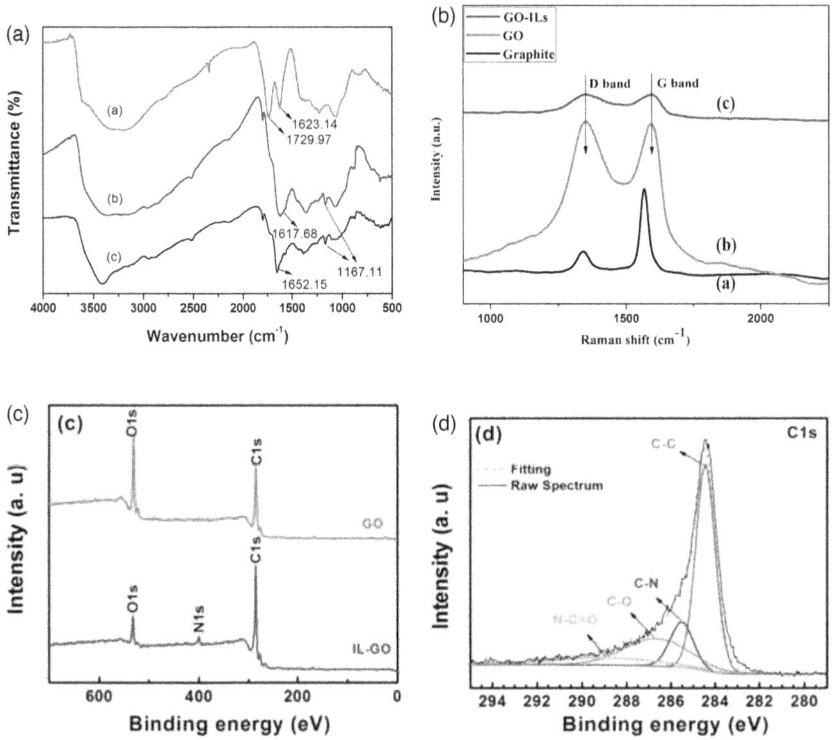

FIGURE 2.7 (a) FTIR spectra of (a) GO, (b) GO-[Aemim][Br], and (c) GO-[Apmim][NTf_2] [40], (b) Raman spectra of (a) graphite, (b) GO, and (c) IL–GO [41], (c) XPS spectra of GO and IL–GO, and (d) XPS C1s spectra of IL–GO [33].

ratio further strengthened the successful grafting of ILs onto the surface of GO, particularly at the edge sites [44, 45].

XPS was used to ascertain the chemical composition and bonding states of GO and IL–GO, as shown in Figure 2.7(c). In the XPS spectrum of GO, different peaks were observed for C1*s* and O1*s*. The IL–GO reported an extra peak for N1*s* at 401.6 eV, confirming the covalent attachment of ILs. In Figure 2.7(d), high-resolution deconvoluted C1*s* spectra for IL–GO revealed four main peaks at 284.5, 285.5, 286.6, and 288.8 eV assigned to the C–C, C–N, C–O, and N–C=O bonds, respectively [46, 47]. The presence of N–C=O bonds supported amide bond formation between IL's amine groups and GO's carboxyl groups [48], while C–N signals confirmed the inclusion of imidazole rings and the occurrence of ring-opening reactions. The carbon-to-oxygen (C/O) ratio significantly increased from 2.29 in GO to 7.87 in IL–GO, indicating a reduction in oxygen-containing groups, aligning with the data of both FTIR and Raman results. Table 2.1 represents the key features, advantages, and disadvantages of different methods employed for synthesizing ILNFs.

TABLE 2.1
Representation of Key Features, Advantages, and Disadvantages of Various Methods Adopted for the Synthesis of ILNFs

Method	Key Features	Advantages	Limitations	References
One-step	Nanoparticles dispersed directly in IL	Simple	Fast stability issues	[23]
Two-step	Nanoparticles synthesized separately and then mixed	Better control	Agglomeration risk	[20]
In-situ polymerization	Polymerization carried out in the presence of IL–nanoparticles mix	Strong interaction	More complex	[30]
Grafting	Nanoparticles surface-modified to bond with ILs	High stability	Time-consuming	[33]

Table 2.2 depicts the various characteristics, such as types of ILs as well as types, concentration, and particle size and shape of the nanoparticles along with study parameters and synthesis methods of the ILNFs.

2.3 ELECTROCHEMICAL PROPERTIES OF ILNFS

2.3.1 Ionic Conductivity and Electrochemical Stability

Although polymers with high dielectric constants are attractive candidates for electrolyte systems, they often suffer from limited ion transport properties. To overcome this drawback, ILs have gained attention as liquid electrolytes due to their superior ionic conductivity and broader electrochemical stability windows [57, 58]. When the nanomaterial favourably interacts with both the IL and the electrode, ILNFs show a wider ESW compared to pure ILs. The [Bmpip][NTf_2] IL, when paired with $LiNTf_2$ lithium salts, displays a broad ESW often exceeding 5–5.5V versus Li/Li^+. This wide potential range enables batteries and capacitors to operate at higher voltages with reduced risk of oxidative or reductive decomposition [49]. ILNFs noticeably enhance electrochemical performance, e.g. the addition of 0.5 wt.% of Au nanoparticles into the [EMIM][$EtSO_4$] IL led to a substantial increase in ionic conductivity of 190% and an improvement in electrolytic capacitance because of the strong attraction between $[EtSO_4]^-$ anions and Au nanoparticles, which change the ionic structure of the IL and modified the charge distribution [59]. Moreover, compared to pure IL, the incorporation of 0.08 wt.% of carbon-encapsulated copper nanoparticles (Cu@C) into [Bmim][PF_6] resulted in a 35 and 65% improvement in diffusion coefficient and electrical conductivity, respectively, which is the result of the interactions between Cu@C nanoparticles and $[Bmim]^+$ cations that not only facilitated ion mobility but also enhanced the ionic bonding energy between $[Bmim]^+$ and $[PF_6]^-$ [60]. This

TABLE 2.2
Summary of Characteristics of ILNFs [49]

Types of ILs	Types of Nanoparticles	Amount of Nanoparticles	Size and Shape of Nanoparticles	Parameters Studied	Method of Synthesis	References
$[C_4mim][NTf_2]$	Al_2O_3	0.5, 1.0, and 2.5 wt.%	50 nm spherical	Density and heat capacity	Vortex mixer	[50]
$[C_6mim][NTf_2]$ $[C_6mim][BF_4]$	SiO_2	0.1, 0.3, 0.5, 1.0, and 2.0 wt.%	10–20 nm spherical	Viscosity and rheology	Ultrasound treatment	[51]
$[C_2mim][DCA]$	MWCNTs, aligned MWCNTs, SWCNTs	1.0 wt.%	MWCNTs: 8 nm diameter, 0.5–2 μm length; aligned MWCNTs: 10–20 nm diameter, 5–15 μm length; SWCNTs: 1–2 nm diameter, 5–30 μm length	Viscosity	Agate mortar	[52]
$[C_2mim][Ac]$	Graphene nanoplatelets	0.05, 0.3, 0.5, 1.0, 2.0, 3.0, and 5.0 mass fraction	5–10 μm diameter, 4–20 nm thickness	Viscosity and thermal conductivity	Ultrasonication	[53]
$[C_2mim][DEP]$	MWCNTs	0.2, 0.5, and 1.0 mass fraction	20–40 nm diameter, 1–2 μm length	Thermal conductivity, viscosity, and density	Magnetic stirrer	[54]
$[C_2mim][DEP]$	SWCNTs	0.5, 1.0, and 2.0 wt.%	1–2 nm diameter, 5–30 μm length	Thermal conductivity	Ultrasonic vibration	[55]
$[C_4mim][NTf_2]$	Al_2O_3	0.5, 1.0, and 2.5 wt.%	50 nm	Density, viscosity, and thermal conductivity	Vortex mixer	[56]
$[C_6mim][BF_4]$	Graphene	0.03 and 0.06 wt.%	Not mentioned	Thermal conductivity, viscosity, specific heat, and density	Ultrasonic vibration	[18]

interaction contributed to increasing the self-diffusion coefficient of the $[Bmim]^+$ cations and reducing the viscosity of the medium, further promoting efficient charge transport. Beyond enhancing conductivity, these nanoparticle-doped ILNFs also displayed superior chemical stability. The formation of stable electric double layers around the nanoparticles minimized nanoparticle corrosion and maintained long-term electrolyte integrity. For example, silver nanoparticles dispersed in imidazolium dicyanamide–based ILs were successfully used in plasmonically enhanced thin DSSC, where they not only improved ionic conductivity but also contributed to the long-term electrochemical stability of the ILNFs [61].

2.3.2 Redox Processes in ILNF Electrolytes

The electrochemical performance of PS/I RFBs is improved by the integration of ILNFs with magnetically modified MWCNTs. Compared to conventional electrolytes, these nanofluids show higher ionic conductivity, faster charge-transfer kinetics, and lower Ohmic and mass-transfer resistances. The incorporation of MWCNTs enhances the energy efficiency of PS/I RFBs to about 80% at 20 mA cm^{-2}, increases peak current densities, and boosts power density by nearly 45%, while maintaining almost 100% Coulombic efficiency with stable cycling for over 200 cycles. This work reports that the electrochemical properties of stable magnetic nanofluidic electrolytes were evaluated using CV on carbon felt electrodes, and their synthesis method was carried out by dispersing MWCNTs into the positive electrolyte at concentrations below 0.3 g l^{-1} [62]. The findings indicate that increasing the MWCNTs content increases the peak current density and the observed values are more uniform. Besides the performance improvement, the intrinsic magnetic properties of MWCNTs permit their easy recovery by magnetic decantation during electrolyte refreshment, which can lead to reduced operating costs over the long term. PS/I RFBs are of considerable significance in efficient energy storage driven by renewables due to the widespread availability of sulphur and low cost as well as the high solubility of polysulphides and iodides. However, conventional PS/I systems suffer disadvantages owing to polysulphide crossover with sulphur precipitation at the positive electrode, low ionic conductivity, slow redox kinetics, and lack of catalytic activity of the carbon-based electrodes that could address some of these problems [63–66]. Earlier procedures employed for modifications of electrodes with various catalysts, including Cu_2XGeS_4, (X = Fe, Co, Ni, Cd, Mn), Cu_7S_4 hollow nanocage catalysis [66] and CoS_2/CoS heterojunction nanoparticles, which have resulted in some improvements but still do not achieve the performance levels noted for state-of-the-market RFB systems [67]. The presence of nanomaterials such as MWCNTs significantly improves the electrochemical surface area and accelerates the reaction kinetics, therefore providing a promising and effective means for pursuing the next-generation PS/I RFB systems.

2.3.2.1 Electrochemical Reactions

The PS/I RFB relies on two key redox couples in aqueous electrolyte, enhanced by ILNFs containing MWCNTs [63, 64]:

At the Positive Electrode (Iodide/Tri-iodide redox):
The primary charge-transfer reactions involve iodide (I^-) and tri-iodide (I_3^-):

$$I_3^- + 2e^- \rightleftharpoons 3I^-, E^0 = +0.5355 V \; vs. \; SHE$$

$$I_2 + 2e^- \rightleftharpoons 2I^-, E^0 = +0.536 V \; vs. \; SHE$$

At Negative Electrode (Polysulphide Redox):

$$S_x^{2-} + 2(x-1)e^- \rightleftharpoons xS^{2-}, E^0 = -0.45 V \; vs. \; SHE$$

In PS/I RFBs, energy storage relies on the iodide/tri-iodide reaction at the positive electrode and PS redox reaction at the negative electrode. The addition of magnetically modified MWCNTs accelerates ion transport and charge transfer in the above reactions, thereby enhancing efficiency, cycling stability, and power density.

2.3.2.2 Role of MWCNTs in ILNF Electrolytes

Enhanced Charge Transfer:
- MWCNTs provide high surface area and conductivity, improving the reaction rate at the carbon felt electrode.
- They reduce charge-transfer resistance in the electrode/electrolyte interface.

Improved Mass Transport:
- Nanotube dispersions help to create a more uniform ionic environment, facilitating diffusion of iodide and polysulphide species.
- This reduces mass-transfer resistance.

Stabilized Electrolyte Conductivity:
- The ILNF shows better ionic conductivity due to the uniform distribution of charged MWCNTs.
- Lower Ohmic resistance (Ω) was observed.

Magnetic Recoverability:
- The magnetic coating allows easy separation of the nanotubes via magnetic decantation, preventing long-term degradation and reducing cost.

2.3.3 Stabilization of Ion

Intermolecular interactions between solute (nanoparticles) and solvent are accountable for the solubility of colloidal nanoparticles in an IL-based solution. Certain IL salts, which exhibit a more intricate network of molecular interactions, are present in liquid form at room temperature. Strong electrostatic (Coulombic) forces and weaker, spatially dependent interactions exist among ILs and solute, playing an important factor to determine the structural behaviour of the nanoparticles. These spatial interactions arise from intermolecular forces between closed-shell molecules. Figure 2.8 illustrates different types of these intermolecular forces [68].

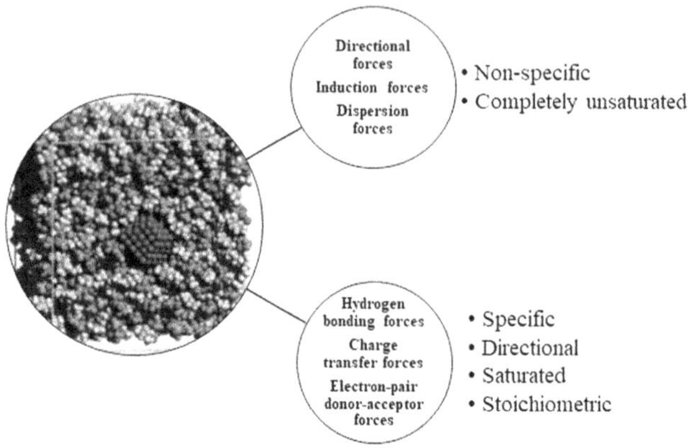

FIGURE 2.8 Schematic diagram of types and characteristics of intermolecular forces [69].

Nanoparticles are stabilized using one of three mechanisms, i.e., electrostatic stabilization, steric stabilization, or a combination of both [70]. The function of stabilizing agent in ILs is to either reduce the strength of attractive Van der Waals interactions or to generate repulsive forces between nanoparticles [71]. These agents form a protective layer around the nanoparticles, effectively preventing their aggregation tendencies. From DLVO theory, the principles of this stabilization process were initially outlined [72, 73]. Figure 2.9 shows the interaction of nanoparticles with ILs and the development of protective electrical double layers (EDLs) resulting from the encapsulation of nanoparticles by ion clusters.

Here are some examples presented in Figure 2.10(a)–(c), demonstrating the formation of EDLs, where charged particles accumulated at the interface between a solid surface and a liquid, resulting in two distinct layers of opposite charges, as well as illustrating how ILs stabilize nanoparticles via EDL formation. From Figure 2.10(a), it can be clearly observed that an inner layer is formed by the negatively

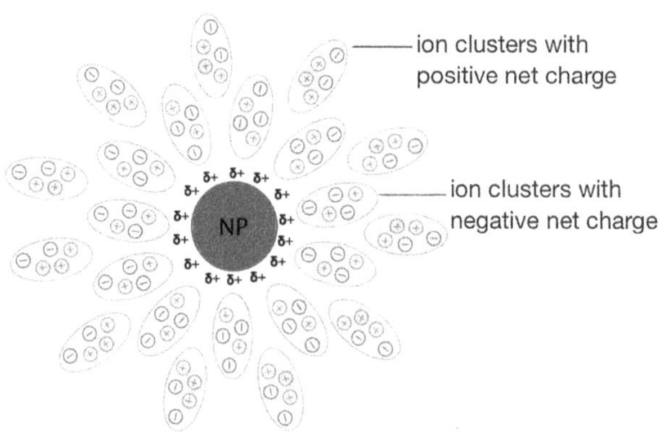

FIGURE 2.9 Schematic representation of the development of EDLs around nanoparticles by the surrounding of ion clusters [74].

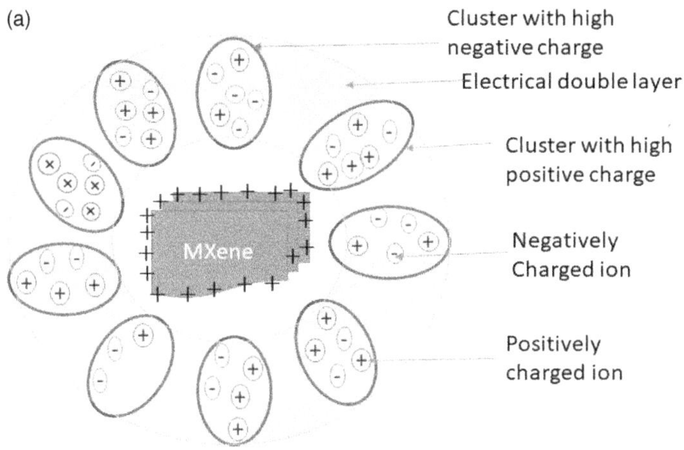

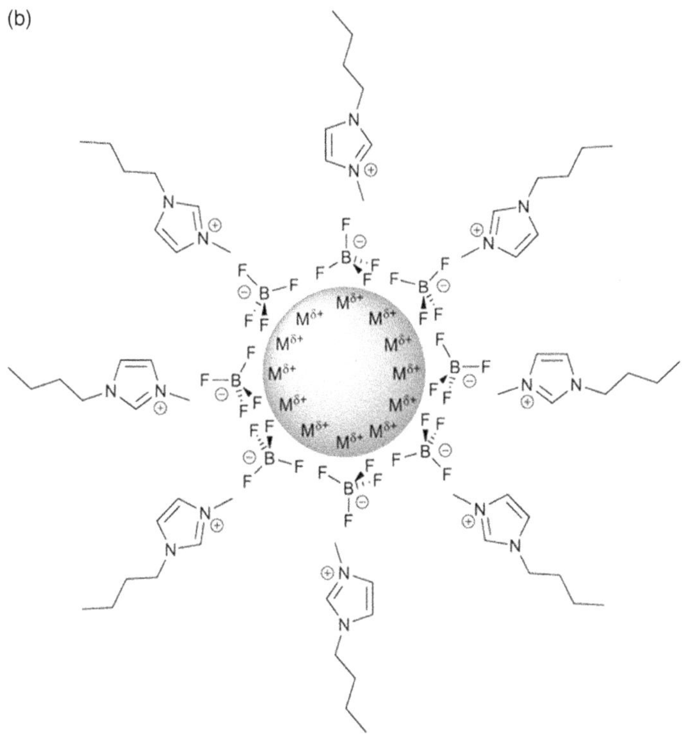

FIGURE 2.10 Schematic illustration of (a) EDL formation around MXene nanoparticles [68], (b) interactions of $[BF_4]^-$ anions and metal nanoparticles with $[Bmim]^+$ cations forming a second EDL shell [76, 78], and (c) electrostatic stabilization of nanoparticles by imidazolium-based ILs [77]. *(Continued)*

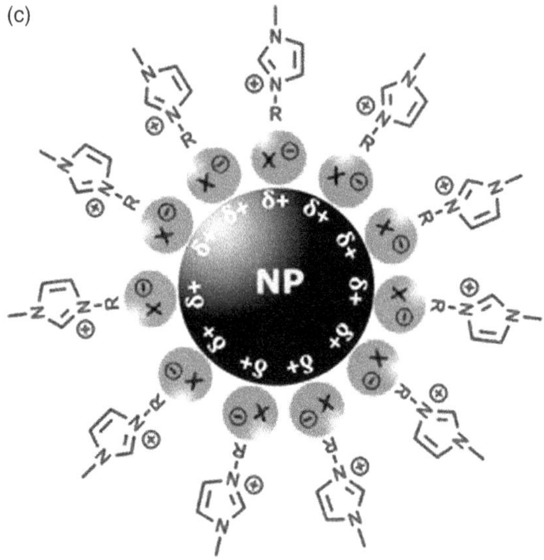

FIGURE 2.10 *(Continued)*

charged surface of MXene nanosheets and positively charged IL. Subsequently, an EDL is formed between this inner layer and an outer layer of IL anions, which prevents aggregation between nanoparticles and ILs. This stabilization mechanism enables ILs to promote electrostatic interactions on nanoparticle surfaces, thereby enhancing their dispersion in the base fluid [75]. Figure 2.10(b) shows a positively charged metal nanoparticle surrounded by IL anions near the surface, with IL cations forming a secondary stabilizing layer that introduces steric hindrance due to the bulky IL groups [76]. Then Figure 2.10(c) generalizes this concept, showing a nanoparticle enveloped by structured layers of IL ions, highlighting both electrostatic and steric effects. Figure 2.10(a)–(c) demonstrates that ILs stabilize nanoparticles by organizing their ions into EDLs that neutralize surface charge and create repulsion between particles. This mechanism applies to various nanoparticle types, regardless of their shape or charge [77].

2.4 APPLICATIONS OF ILNFs

2.4.1 Thermal Conductivity

Nanoparticles can dramatically enhance the thermal conductivity of base fluids even at low concentrations (0.001–5% w/v), with CNT-based nanofluids showing improvements up to 161% [79]. Murshed *et al.* introduced ILNFs by dispersing MWCNTs in ILs such as [Emim][$EtSO_4$] and [Bmim][Tf_2N] [80]. Classical models such as Maxwell (1881) [81], Hamilton–Crosser (1962) [82], and Bruggeman (1935) [83] remain relevant for modelling ILNFs because they provide the fundamental theoretical frameworks for predicting effective thermal conductivity in heterogeneous systems. These models are still widely used as reference points and benchmarks when

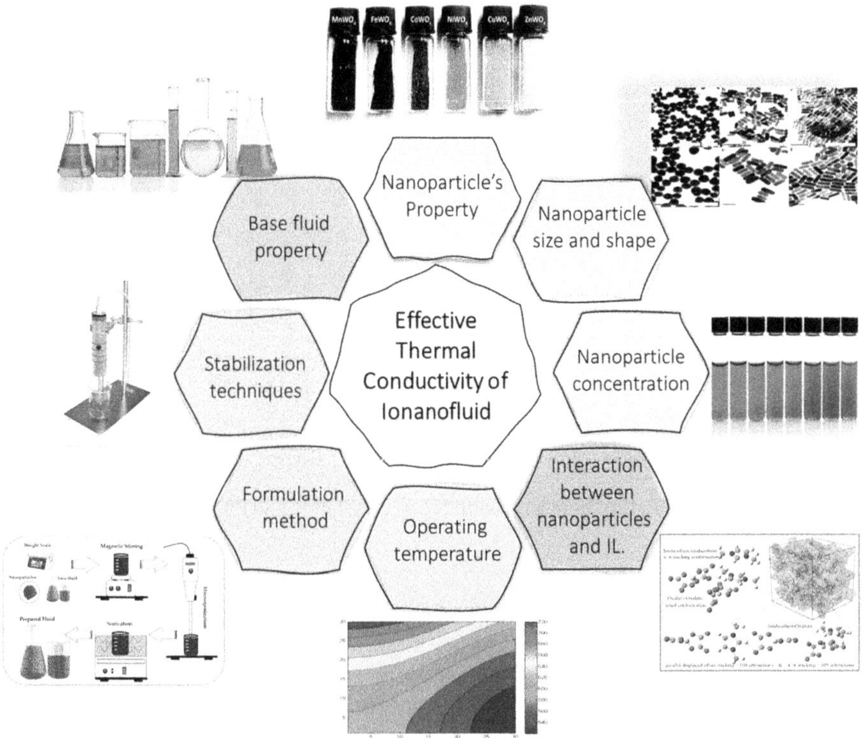

FIGURE 2.11 Different factors affecting the thermal conductivity of ILNFs [3].

developing or validating advanced correlations. However, they oversimplify particle interactions, assume uniform particle shapes, and neglect interfacial phenomena, which limits their accuracy in predicting complex fluids. Modern studies demonstrate that the thermal conductivity of ILNFs is not governed solely by particle loading but is strongly influenced by interfacial layering, nanoparticle morphology, Brownian motion, and processing parameters such as p^H and sonication. As a result, these classical models are being modified to better predict and tailor the thermal behaviour of advanced ILNFs. Figure 2.11 describes various factors affecting the thermal conductivity of the ILNFs.

2.4.2 Redox Flow Batteries (RFBs)

RFBs are emerging energy storage systems that rely on external liquid electrolytes undergoing reversible redox reactions. Incorporating ILNFs into these electrolytes has been shown to substantially enhance performance by improving conductivity, electrochemical activity, and energy density. For example, graphene, *r*GO, and magnetic CNTs have all been tested with promising results [84]. A study using magnetic-modified MWCNTs in a PS/I flow battery demonstrated that even a low concentration (0.3 g l^{-1}) boosted power density by 45% and achieved nearly

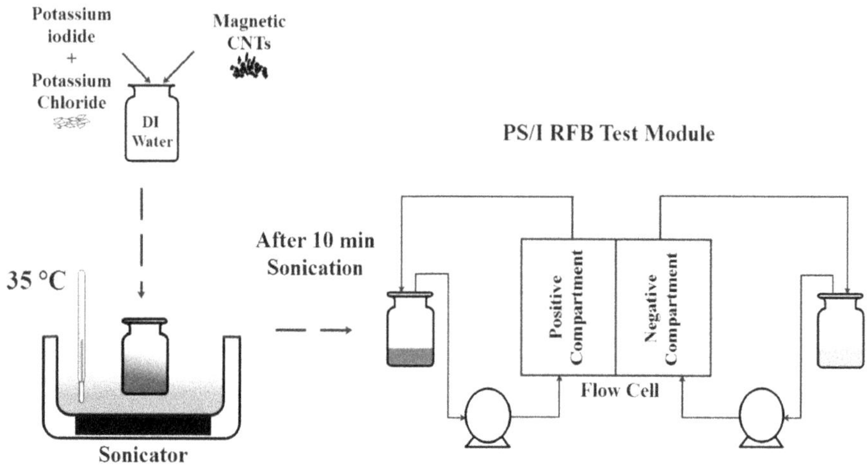

FIGURE 2.12 Synthesis of magnetic nanofluidic electrolyte and PS/I RFB test module [62].

80% energy efficiency [62]. Similarly, Lobato *et al.* [85] showed that the reduction of Ohmic resistance and expansion of the active surface area for redox reactions can be achieved by increasing nanocarbon concentration in vanadium-based nanofluids. From the above studies, it can be concluded that the concentration and interfacial effects on nanoparticles increase the performance of ILNFs in RFBs. When there is an enhancement in conductivity and surface activity, a trade-off occurs with viscosity and long-term stability; therefore, optimization strategies are required. The synthesis of magnetic nanofluidic electrolytes and their evaluation in an RFB setup are shown in Figure 2.12. In Figure 2.12, KCl acts as a supporting electrolyte to improve ionic conductivity and stability, while KI functions as the redox-active species through the iodide/tri-iodide couple. Due to sonication, uniform dispersion of magnetic CNTs takes place, which enhances the charge-transfer property. The iodide/tri-iodide (positive side) and polysulphide (negative side) couples facilitate the redox reactions, and the prepared electrolyte is circulated within the flow cell.

2.4.3 Dye-Sensitized Solar Cells (DSSCs)

In the DSSCs, electrons in the dye molecules are excited by sunlight and flow through a semiconductor to an external circuit, generating electricity, as represented in Figure 2.13. The oxidized dye is regenerated by a redox mediator in the electrolyte, which is itself regenerated at the counter electrode, enhanced here by MWCNT films. To increase DSSC efficiency, ILNFs facilitate faster ion transport and stabilize the electrolyte. Due to weaker ionic bonding and reduced viscosity, Cu nanoparticles in [Bmim][PF_6] increased the electrical conductivity by 65% and the diffusion coefficient by 35% [86]. MWCNTs in [Pmim][I] formed hydrogen bonds with cations, enhancing ion mobility and lowering Coulombic attraction, as

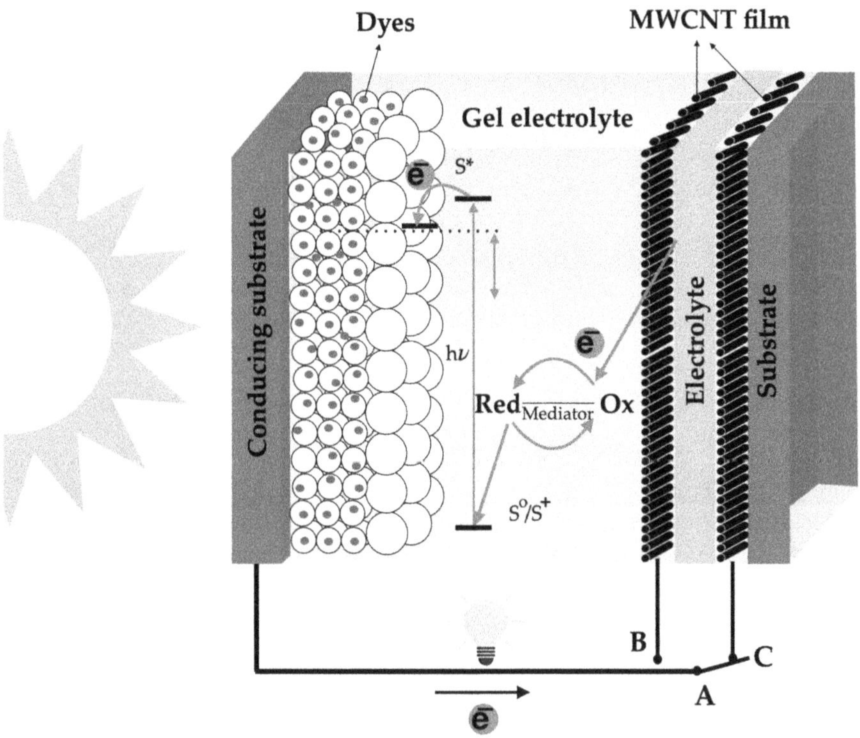

FIGURE 2.13 Schematic representation of DSSCs [57].

reported by Neo and Ouyang [12]. Correspondingly, Adhyaksa *et al.* [61] explained that Ag nanoparticles in imidazolium-dicyanamide IL were protected from corrosion by enhancing device durability and ionic double layer formation. Generally, these results indicate that additional conduction paths are created and aggregation of ions is disordered by nanoparticles, while the predominance of mobility or stability is determined by specific ion–nanoparticle interactions. Future DSSC design can be focused on the optimization of IL–nanoparticle interactions to achieve a balance between long-term stability and conductivity. In these systems, ILs and nanoparticles synergistically act to improve overall cell stability, reduce viscosity, and enhance conductivity.

2.4.4 Fuel Desulfurization Processes (FDP)

Figure 2.14 demonstrates how ILNFs enhance the desulfurization process compared to pure ILs. To improve extraction efficiency, TiO_2 and Fe_2O_3 nanoparticles were incorporated for the first time into pure ILs in the form of [Dmim][MP], [Emim][SCN], and [Bmim][SCN]. Among the ILs tested, [Bmim][SCN] exhibited the best performance. The triangular diagram presents the LLE data, revealing Type 2 phase behaviour, which indicates the effective partitioning of thiophene into the IL phase.

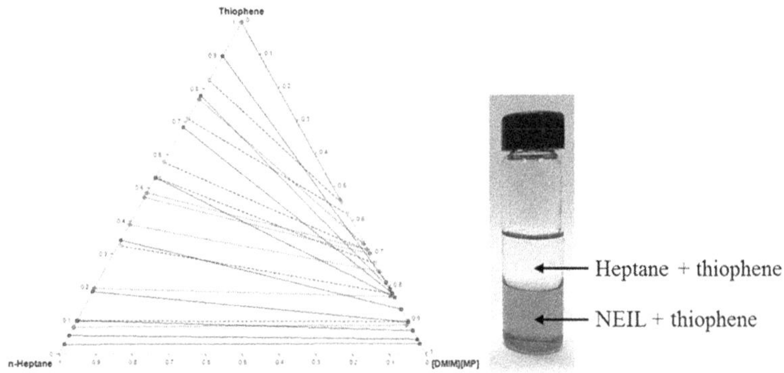

FIGURE 2.14 Tie line of ternary mixtures [n-heptane + thiophene + ILNFs] [87].

Both the distribution ratio and selectivity significantly increased in the presence of nanoparticles. Interestingly, ILs with longer alkyl chains improved distribution but reduced selectivity, suggesting a structure–property relationship. These findings highlight that the nanoparticles not only improve partitioning efficiency but also alter phase behaviour, offering a tunable platform for fuel purification using ILNFs [87].

2.4.5 Supercapacitor Application

ILNFs represent one of the most promising classes of electrolytes for high-performance supercapacitors, as they combine the wide ESW (3–6 V) of ILs with the enhanced ionic conductivity provided by nanoparticles. This synergistic effect accelerates charge migration, reduces internal resistance, and supports higher energy and power densities ($E \propto V^2$) [49]. This synergistic combination of high ionic conductivity and broad stability range allows ILNF-based supercapacitors to achieve superior rate capability, higher energy and power densities, and excellent cycling stability compared to devices using pure ILs.

The CV curves display equal shapes at different scan rates, indicating superior electrochemical stability, as shown in Figure 2.15(a) and (b). Similarly, Figures 2.15(c) and (d) represent the GCD plots at various current densities. All GCD diagrams show nearly linear curves, reflecting excellent electrochemical performance. For instance, RGO-IL1 ([Bmim][BF_4]) and RGO-IL2 ([Bmim][PF_6]) supercapacitors achieved maximum energy densities of 32.02 and 50.19 Wh kg^{-1}, respectively, with both maintaining strong performance even at high power densities. Importantly, RGO-IL2, despite having only a moderate voltage window (3.0 V), outperformed many graphene-based supercapacitors, delivering the highest energy density reported so far. This can be attributed to the interplay between ionic conductivity and ESW. Enhanced conductivity minimizes polarization losses, while a stable ESW prevents side reactions, together enabling long cycle life and high energy storage efficiency. A practical demonstration of this capability is shown in Figure 2.15(f), where a single RGO-IL2 device successfully powered an LED (with a capacity of 18 pF and a rated voltage of 2.4 V) for 30 s.

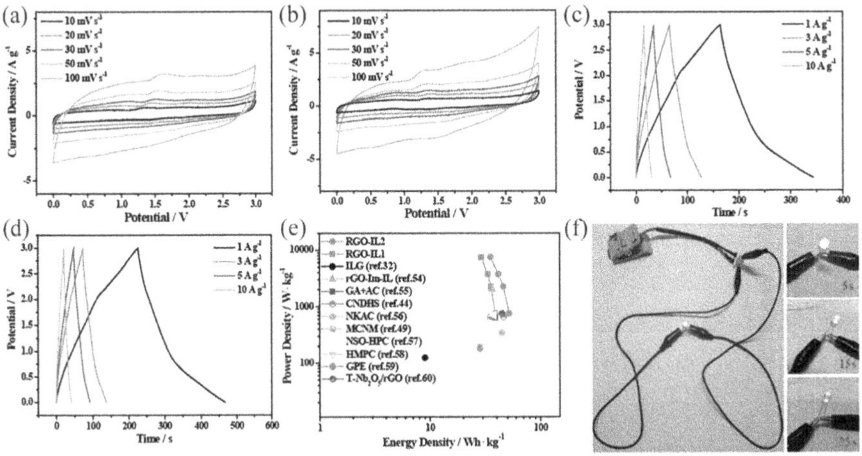

FIGURE 2.15 (a and b) CV curves of RGO-IL1 and RGO-IL2 at different scan rates, (c and d) GCD curves of RGO-IL1 and RGO-IL2 at various current densities in [Bmim]PF6/AN (1:1), (e) Ragone plots of symmetric supercapacitors based on graphene, and (f) photograph of an LED powered by RGO-IL2 symmetric supercapacitors [88].

2.5 CONCLUSION AND FUTURE SCOPE

ILNFs are an innovative category of engineered fluids formed by incorporating nanoparticles such as carbon nanomaterials or metal/metal oxide nanoparticles into the ILs. These ILs are the organic salts, which remain liquid at or around normal temperature, serve as the base fluid, and contribute unique physicochemical characteristics. The combination results in a synergistic enhancement of properties, making ILNFs highly effective in thermal and energy-related applications. ILNFs show excellent thermal stability, low vapour pressure, superior heat transfer capability, and high resistance to combustion, outperforming conventional nanofluids that typically use glycol, oil, or water as their base fluids. Better stabilization and dispersion of nanoparticles are due to the ionic nature of the liquid medium, which significantly minimize agglomeration and sedimentation over time. Moreover, the customizable molecular structure of ILs allows researchers to fine-tune their viscosity, polarity, and thermal behaviour to meet specific performance needs. High-temperature environments are suitable for ILNFs due to their improved characteristics. These high-temperature environments include industrial heat exchangers, advanced electronics cooling, nuclear reactors, and solar thermal systems. Beyond thermal management, ILNFs offer multifunctional advantages that can also function as electrochemical media and lubricants, making them applicable in fields like tribology, microfluidics, and energy storage devices such as batteries and supercapacitors.

Despite their many advantages, ILNFs face notable limitations that may hinder large-scale adoption. The relatively high cost of ILs, potential eco-toxicity, and uncertainties regarding their environmental footprint pose significant barriers to commercialization. Moreover, scaling laboratory synthesis methods into industrially viable and reproducible processes remains a critical challenge. A deeper molecular-level understanding of IL–nanoparticle interactions is essential to optimize stability and long-term performance. Looking forward, research should focus on designing eco-friendly and cost-effective ILs, developing scalable and sustainable synthesis routes, and functionalizing nanoparticles to ensure robust compatibility. The integration of advanced computational modelling with experimental validation will further accelerate insights into heat and mass-transfer mechanisms. Exploring hybrid ILNFs, which incorporate diverse nanoparticles, may unlock multifunctional properties that extend beyond thermal management. In conclusion, ILNFs represent a transformative class of engineered fluids with immense potential. Still, their successful transition from laboratory innovation to industrial application will depend on addressing environmental safety, scalability, and economic feasibility through sustained interdisciplinary research.

ABBREVIATIONS

[Apmim][NTf_2]	1-aminopropyl-3-methylimidazolium bis(trifluoromethylsulfonyl)imine
[Dmim][MP]	1,3-dimethylimidazolium methylphosphonate
[Aemim][Br]	1-aminoethyl-3-methylimidazolium bromide
[Bmpip][NTf_2]	1-butyl-1-methylpiperidinium bis(trifluoromethylsulfonyl) imide
[Bmim][PF_6]	1-butyl-3-methylimidazolium hexafluorophosphate
[Bmim][SCN]	1-butyl-3-methylimidazolium thiocyanate
[Cmmim][Cl]	1-carboxymethyl-3-methylimidazolium chloride
[EMIM][$EtSO_4$]	1-ethyl-3-methylimidazolium ethyl sulphate
[C_2mim][CH_3SO_3]	1-ethyl-3-methylimidazolium methanesulfonate
[Emim][SCN]	1-ethyl-3-methylimidazolium thiocyanate
[Pmim][I]	1-propyl-3-methylimidazolium iodide
APTES	3-aminopropyltriethoxysilane
CNT	carbon nanotube
CSP	concentrated solar power
CV	cyclic voltammetry
DLVO	Derjaguin–Landau–Verwey–Overbeek
DSSCs	dye-sensitized solar cells
DLS	dynamic light scattering
EDLs	electrical double layers
ESW	electrochemical stability window
FDP	fuel desulfurization processes
GCD	galvanostatic charge–discharge
AuNFs	gold nanoframes
GO	graphene oxide
ILNFs	ionanofluids
ILs	ionic liquids
LED	light-emitting diode
LLE	liquid-liquid equilibrium

$LiNTf_2$	lithium bis(trifluoromethanesulfonyl)imide
MNPs	magnetic nanoparticles
MWCNTs	multi-walled carbon nanotubes
NEILs	nanoparticle-enhanced ILs
NSAIDs	nonsteroidal anti-inflammatory drugs
PILs	polymeric ILs
PS/I	polysulphide–iodide
PVP	polyvinylpyrrolidone
RFBs	redox flow batteries
***r*GO**	reduced graphene oxide
SEM	scanning electron microscopy
SWCNTs	single-walled carbon nanotubes
TA	tannic acid
TEM	transmission electron microscopy
XPS	X ray photoelectron spectroscopy

REFERENCES

1. Joseph, A., Fal, J., Bąk, A., Mathew, S. and Żyła, G., 2023. Complex dielectric response and EDL characteristics of different types of ionic liquid iron oxide nanofluids (ionanofluids). *Chemical Physics Impact*, *7*, p.100357. https://doi.org/10.1016/j.chphi.2023.100357
2. Mohapatra, P. and Barick, A.K., 2025. Ionic liquids based polymer electrolytes for supercapacitor applications. *Journal of Power Sources*, *626*, p.235749. https://doi.org/10.1016/j.jpowsour.2024.235749
3. Das, L., Rubbi, F., Habib, K., Aslfattahi, N., Saidur, R., Saha, B.B., Algarni, S., Irshad, K. and Alqahtani, T., 2021. State-of-the-art ionic liquid and ionanofluids incorporated with advanced nanomaterials for solar energy applications. *Journal of Molecular Liquids*, *336*, p.116563. https://doi.org/10.1016/j.molliq.2021.116563
4. Nieto de Castro, C.A., Lourenço, M.J.V., Ribeiro, A.P.C., Langa, E., Vieira, S.I.C., Goodrich, P. and Hardacre, C., 2010. Thermal properties of ionic liquids and ionanofluids of imidazolium and pyrrolidinium liquids. *Journal of Chemical and Engineering Data*, *55*(2), pp.653–661. https://doi.org/10.1021/je900648p
5. Main, K.L., Eberl, B.K., McDaniel, D., Tikadar, A., Paul, T.C. and Khan, J.A., 2021. Nanoparticles size effect on thermophysical properties of ionic liquids based nanofluids. *Journal of Molecular Liquids*, *343*, p.117609. https://doi.org/10.1016/j.molliq.2021.117609
6. Choi, S.U. and Eastman, J.A., 1995. Enhancing Thermal Conductivity of Fluids with Nanoparticles. *ASME International Mechanical Engineering Congress and Exposition*, 12–17 November 1995, San Francisco, CA, USA. https://www.osti.gov/biblio/196525
7. Chereches, E.I., Bejan, D., Ibanescu, C., Danu, M. and Minea, A.A., 2021. Ionanofluids with $[C_2mim][CH_3SO_3]$ ionic liquid and alumina nanoparticles: an experimental study on viscosity, specific heat and electrical conductivity. *Chemical Engineering Science*, *229*, p.116140. https://doi.org/10.1016/j.ces.2020.116140
8. Ferreira, A.G.M., Simões, P.N., Ferreira, A.F., Fonseca, M.A., Oliveira, M.S.A. and Trino, A.S.M., 2013. Transport and thermal properties of quaternary phosphonium ionic liquids and ionanofluids. *The Journal of Chemical Thermodynamics*, *64*, pp.80–92. https://doi.org/10.1016/j.jct.2013.04.013

9. Liavitskaya, T., Paulechka, E., Blokhin, A.V. and Shevelyova, M., 2019. Thermodynamic behaviour and polymorphism of 1-butyl-3-methylimidazolium hexafluorophosphate composites with multi-walled carbon nanotubes. *The Journal of Chemical Thermodynamics*, *131*, pp.262–268. https://doi.org/10.1016/j.jct.2018.11.006
10. França, J.M., Lourenço, M.J.V., Murshed, S.S., Pádua, A.A. and Nieto de Castro, C.A., 2018. Thermal conductivity of ionic liquids and ionanofluids and their feasibility as heat transfer fluids. *Industrial and Engineering Chemistry Research*, *57*(18), pp.6516–6529. https://doi.org/10.1021/acs.iecr.7b04770
11. Shevelyova, M.P., Paulechka, Y.U., Kabo, G.J., Blokhin, A.V., Kabo, A.G. and Gubarevich, T.M., 2013. Physicochemical properties of imidazolium-based ionic nanofluids: density, heat capacity, and enthalpy of formation. *The Journal of Physical Chemistry C*, *117*(9), pp.4782–4790. https://doi.org/10.1021/jp3059432
12. Neo, C.Y. and Ouyang, J., 2012. Functionalized carbon nanotube-induced viscosity reduction of an ionic liquid and performance improvement of dye-sensitized solar cells. *Electrochimica Acta*, *85*, pp.1–8. https://doi.org/10.1016/j.electacta.2012.08.041
13. Alizadeh, J. and Moraveji, M.K., 2018. An experimental evaluation on thermophysical properties of functionalized graphene nanoplatelets ionanofluids. *International Communications in Heat and Mass Transfer*, *98*, pp.31–40. https://doi.org/10.1016/j.icheatmasstransfer.2018.08.008
14. Chandra, A., Sistla, Y.S., Ahmed, M.A., Vasireddy, D.V.S., Jaglan, N., Das, N.K., Banerjee, T. and Sistla, V.S., 2025. Heat transfer fluids: amino acid anion ionic liquid based IoNanofluids with remarkable thermal conductivity and low viscosity. *RSC Advances*, *15*(29), pp.23146–23164.
15. Hosseinghorbani, A., Mozaffarian, M. and Pazuki, G., 2020. Application of graphene oxide ionanofluid as a superior heat transfer fluid in concentrated solar power plants. *International Communications in Heat and Mass Transfer*, *111*, p.104450. https://doi.org/10.1016/j.icheatmasstransfer.2019.104450
16. Chen, W., Zou, C. and Li, X., 2017. An investigation into the thermophysical and optical properties of SiC/ionic liquid nanofluid for direct absorption solar collector. *Solar Energy Materials and Solar Cells*, *163*, pp.157–163. https://doi.org/10.1016/j.solmat.2017.01.029
17. Liu, J., Xu, C., Chen, L., Fang, X. and Zhang, Z., 2017. Preparation and photo-thermal conversion performance of modified graphene/ionic liquid nanofluids with excellent dispersion stability. *Solar Energy Materials and Solar Cells*, *170*, pp.219–232. https://doi.org/10.1016/j.solmat.2017.05.062
18. Liu, J., Wang, F., Zhang, L., Fang, X. and Zhang, Z., 2014. Thermodynamic properties and thermal stability of ionic liquid-based nanofluids containing graphene as advanced heat transfer fluids for medium-to-high-temperature applications. *Renewable Energy*, *63*, pp.519–523. https://doi.org/10.1016/j.renene.2013.10.002
19. Chereches, E.I., Prado, J.I., Chereches, M., Minea, A.A. and Lugo, L., 2019. Experimental study on thermophysical properties of alumina nanoparticle enhanced ionic liquids. *Journal of Molecular Liquids*, *291*, p.111332. https://doi.org/10.1016/j.molliq.2019.111332
20. Minea, A.A. and Sohel Murshed, S.M., 2021. Ionic liquids-based nanocolloids—A review of progress and prospects in convective heat transfer applications. *Nanomaterials*, *11*(4), p.1039. https://doi.org/10.3390/nano11041039
21. Wen, D. and Ding, Y., 2005. Formulation of nanofluids for natural convective heat transfer applications. *International Journal of Heat and Fluid Flow*, *26*(6), pp.855–864. https://doi.org/10.1016/j.ijheatfluidflow.2005.10.005
22. Wen, D., Ding, Y. and Williams, R.A., 2005. Pool boiling heat transfer of aqueous based TiO_2 nanofluids. *Journal of Enhanced Heat Transfer*, *13*(3), pp.245–260.

23. Zhang, H., Cui, H., Yao, S., Zhang, K., Tao, H. and Meng, H., 2012. Ionic liquid-stabilized non-spherical gold nanofluids synthesized using a one-step method. *Nanoscale Research Letters*, *7*, p.583. https://doi.org/10.1186/1556-276X-7-583
24. Zhang, H. and Cui, H., 2009. Synthesis and characterization of functionalized ionic liquid-stabilized metal (gold and platinum) nanoparticles and metal nanoparticle/carbon nanotube hybrids. *Langmuir*, *25*(5), pp.2604–2612. https://doi.org/10.1021/la803347h
25. Newman, J.D.S. and Blanchard, G.J., 2006. Formation of gold nanoparticles using amine reducing agents. *Langmuir*, *22*(13), pp.5882–5887. https://doi.org/10.1021/la060045z
26. Minea, A.A. and Chercheş, E.I., 2024. Experimental studies on thermal conductivity and heat transfer of 1-Butyl-3-methylimidazolium tetrafluoroborate ionic liquid and its nanocolloids. *International Communications in Heat and Mass Transfer*, *154*, p.107406. https://doi.org/10.1016/j.icheatmasstransfer.2024.107406
27. Murshed, S.M.S. and de Castro, C.A.N., 2014. *Nanofluids: synthesis, properties, and applications*. Nova Science Publishers, Hauppauge, NY.
28. de Castro, C.A.N., Vieira, S.I., Lourenco, M.J. and Murshed, S.M., 2017. Understanding stability, measurements, and mechanisms of thermal conductivity of nanofluids. *Journal of Nanofluids*, *6*(5), pp.804–811. https://doi.org/10.1166/jon.2017.1388
29. Huminic, G. and Huminic, A., 2021. Heat transfer capability of ionanofluids for heat transfer applications. *International Journal of Thermophysics*, *42*(12), pp.1–17. https://doi.org/10.1007/s10765-020-02757-x
30. Lindner, J.P., 2016. Imidazolium-based polymers via the poly-Radziszewski reaction. *Macromolecules*, *49*(6), pp.2046–2053. https://doi.org/10.1021/acs.macromol.5b02417
31. Castro-Grijalba, A., Reyes-Gallardo, E.M., Wuilloud, R.G., Lucena, R. and Cárdenas, S., 2017. Synthesis of magnetic polymeric ionic liquid nanocomposites by the Radziszewski reaction. *RSC Advances*, *7*(68), pp.42979–42985. https://doi.org/10.1039/C7RA07947H
32. Yuan, C., Zhang, J., Chen, G. and Yang, J., 2011. Insight into carbon nanotube effect on polymer molecular orientation: an infrared dichroism study. *Chemical Communications*, *47*(3), pp.899–901. https://doi.org/10.1039/C0CC03198D
33. Liu, C., Qiu, S., Du, P., Zhao, H. and Wang, L., 2018. An ionic liquid–graphene oxide hybrid nanomaterial: synthesis and anticorrosive applications. *Nanoscale*, *10*(17), pp.8115–8124. https://doi.org/10.1039/C8NR01890A
34. Shubhadarshinee, L., Jali, B.R., Barick, A.K. and Mohapatra, P., 2022. Preparation and characterisation of silver nanoparticles/graphene oxide hybrid nanofiller reinforced-polyaniline. *Plastics, Rubber and Composites: Macromolecular Engineering*, *51*(2), pp.72–84. https://doi.org/10.1080/14658011.2021.1939587
35. Moghadam, M.H.M., Sabury, S., Gudarzi, M.M. and Sharif, F., 2014. Graphene oxide-induced polymerization and crystallization to produce highly conductive polyaniline/graphene oxide composite. *Journal of Polymer Science Part A: Polymer Chemistry*, *52*(11), pp.1545–1554. https://doi.org/10.1002/pola.27147
36. Xu, J., Wang, L. and Zhu, Y., 2012. Decontamination of bisphenol A from aqueous solution by graphene adsorption. *Langmuir*, *28*(22), pp.8418–8425. https://doi.org/10.1021/la301476p
37. Chandra, V., Park, J., Chun, Y., Lee, J.W., Hwang, I.C. and Kim, K.S., 2010. Water-dispersible magnetite-reduced graphene oxide composites for arsenic removal. *ACS Nano*, *4*(7), pp.3979–3986. https://doi.org/10.1021/nn1008897
38. Yang, H., Shan, C., Li, F., Han, D., Zhang, Q. and Niu, L., 2009. Covalent functionalization of polydisperse chemically-converted graphene sheets with amine-terminated ionic liquid. *Chemical Communications*, (26), pp.3880–3882. https://doi.org/10.1039/B905085J

39. Ding, X., Wang, Y., Wang, Y., Pan, Q., Chen, J., Huang, Y. and Xu, K., 2015. Preparation of magnetic chitosan and graphene oxide-functional guanidinium ionic liquid composite for the solid-phase extraction of protein. *Analytica Chimica Acta*, *861*, pp.36–46. https://doi.org/10.1016/j.aca.2015.01.004
40. Zhou, X., Zhang, Y., Huang, Z., Lu, D., Zhu, A. and Shi, G., 2016. Ionic liquids modified graphene oxide composites: a high efficient adsorbent for phthalates from aqueous solution. *Scientific Reports*, *6*, p.38417. https://doi.org/10.1038/srep38417
41. Das, M., Aswathy, T.R., Pal, S. and Naskar, K., 2021. Effect of ionic liquid modified graphene oxide on mechanical and self-healing application of an ionic elastomer. *European Polymer Journal*, *158*, p.110691. https://doi.org/10.1016/j.eurpolymj.2021.110691
42. Kudin, K.N., Ozbas, B., Schniepp, H.C., Prud'Homme, R.K., Aksay, I.A. and Car, R., 2008. Raman spectra of graphite oxide and functionalized graphene sheets. *Nano Letters*, *8*(1), pp.36–41. https://doi.org/10.1021/nl071822y
43. Park, S., Dikin, D.A., Nguyen, S.T. and Ruoff, R.S., 2009. Graphene oxide sheets chemically cross-linked by polyallylamine. *The Journal of Physical Chemistry C*, *113*(36), pp.15801–15804. https://doi.org/10.1021/jp907613s
44. Song, B., Sizemore, C., Li, L., Huang, X., Lin, Z., Moon, K.S. and Wong, C.P., 2015. Triethanolamine functionalized graphene-based composites for high performance supercapacitors. *Journal of Materials Chemistry A*, *3*(43), pp.21789–21796. https://doi.org/10.1039/C5TA05674H
45. Ryu, S.H. and Shanmugharaj, A.M., 2014. Influence of long-chain alkylamine-modified graphene oxide on the crystallization, mechanical and electrical properties of isotactic polypropylene nanocomposites. *Chemical Engineering Journal*, *244*, pp.552–560. https://doi.org/10.1016/j.cej.2014.01.101
46. Huang, Y.L., Tien, H.W., Ma, C.C.M., Yang, S.Y., Wu, S.Y., Liu, H.Y. and Mai, Y.W., 2011. Effect of extended polymer chains on properties of transparent graphene nanosheets conductive film. *Journal of Materials Chemistry*, *21*(45), pp.18236–18241. https://doi.org/10.1039/C1JM13790E
47. Li, Z., Wang, R., Young, R.J., Deng, L., Yang, F., Hao, L., Jiao, W. and Liu, W., 2013. Control of the functionality of graphene oxide for its application in epoxy nanocomposites. *Polymer*, *54*(23), pp.6437–6446. https://doi.org/10.1016/j.polymer.2013.09.054
48. Sun, Y., Fang, Z., Wang, C., Ariyawansha, K.R.M., Zhou, A. and Duan, H., 2015. Sandwich-structured nanohybrid paper based on controllable growth of nanostructured MnO_2 on ionic liquid functionalized graphene paper as a flexible supercapacitor electrode. *Nanoscale*, *7*(17), pp.7790–7801. https://doi.org/10.1039/C5NR00946D
49. Paul, T.C., Tikadar, A., Mahamud, R., Salman, A.S., Morshed, A.M. and Khan, J.A., 2021. A critical review on the development of ionic liquids-based nanofluids as heat transfer fluids for solar thermal energy. *Processes*, *9*(5), p.858. https://doi.org/10.3390/pr9050858
50. Bridges, N.J., Visser, A.E. and Fox, E.B., 2011. Potential of nanoparticle-enhanced ionic liquids (NEILs) as advanced heat-transfer fluids. *Energy and Fuels*, *25*(10), pp.4862–4864. https://doi.org/10.1021/ef2012084
51. Wittmar, A., Ruiz-Abad, D. and Ulbricht, M., 2012. Dispersions of silica nanoparticles in ionic liquids investigated with advanced rheology. *Journal of Nanoparticle Research*, *14*, p.651. https://doi.org/10.1007/s11051-011-0651-1
52. Pamies, R., Avilés, M.D., Arias-Pardilla, J., Carrión, F.J., Sanes, J. and Bermúdez, M.D., 2019. Rheological study of new dispersions of carbon nanotubes in the ionic liquid 1-ethyl-3-methylimidazolium dicyanamide. *Journal of Molecular Liquids*, *278*, pp.368–375. https://doi.org/10.1016/j.molliq.2019.01.074
53. Zhang, F.F., Zheng, F.F., Wu, X.H., Yin, Y.L. and Chen, G., 2019. Variations of thermophysical properties and heat transfer performance of nanoparticle-enhanced ionic liquids. *Royal Society Open Science*, *6*(4), p.182040. http://dx.doi.org/10.1098/rsos.182040

54. Xie, H., Zhao, Z., Zhao, J. and Gao, H., 2016. Measurement of thermal conductivity, viscosity and density of ionic liquid [EMIM][DEP]-based nanofluids. *Chinese Journal of Chemical Engineering*, *24*(3), pp.331–338. https://doi.org/10.1016/j.cjche.2015.11.024
55. Li, C., Zhao, Z., Zhang, X. and Li, T., 2018. Simulation and experimental study on thermal conductivity of [EMIM][DEP]+H_2O+SWCNTs nanofluids as a new working pairs. *International Journal of Thermophysics*, *39*, p.41. https://doi.org/10.1007/s10765-018-2359-2
56. Paul, T.C., Morshed, A.K.M.M., Fox, E.B. and Khan, J.A., 2015. Thermal performance of Al_2O_3 nanoparticle enhanced ionic liquids (NEILs) for concentrated solar power (CSP) applications. *International Journal of Heat and Mass Transfer*, *85*, pp.585–594. https://doi.org/10.1016/j.ijheatmasstransfer.2015.01.071
57. Pereira, J., Souza, R., Moita, A. and Moreira, A., 2023. Nanofluids and ionic fluids as liquid electrodes: an overview on their properties and potential applications. *Processes*, *11*(11), p.3189. https://doi.org/10.3390/pr11113189
58. Galiński, M., Lewandowski, A. and Stępniak, I., 2006. Ionic liquids as electrolytes. *Electrochimica acta*, *51*(26), pp.5567–5580. https://doi.org/10.1016/j.electacta.2006.03.016
59. Cheng, F., Liang, J., Tao, Z. and Chen, J., 2011. Functional materials for rechargeable batteries. *Advanced Materials*, *23*(15), pp.1695–1715. https://doi.org/10.1002/adma.201003587
60. Hamm, S.C., Basuray, S., Mukherjee, S., Sengupta, S., Mathai, J.C., Baker, G.A. and Gangopadhyay, S., 2014. Ionic conductivity enhancement of sputtered gold nanoparticle-in-ionic liquid electrolytes. *Journal of Materials Chemistry A*, *2*(3), pp.792–803. https://doi.org/10.1039/C3TA13431H
61. Adhyaksa, G.W.P., Baek, S.W., Lee, G.I., Lee, D.K., Lee, J.Y. and Kang, J.K., 2014. Coupled near-and far-field scattering in silver nanoparticles for high-efficiency, stable, and thin plasmonic dye-sensitized solar cells. *ChemSusChem*, *7*(9), pp.2461–2468. https://doi.org/10.1002/cssc.201402146
62. Rahimi, M., Dehkordi, A.M. and Roberts, E.P., 2021. Magnetic nanofluidic electrolyte for enhancing the performance of polysulfide/iodide redox flow batteries. *Electrochimica Acta*, *369*, p.137687. https://doi.org/10.1016/j.electacta.2020.137687
63. Li, Z., Weng, G., Zou, Q., Cong, G. and Lu, Y.C., 2016. A high-energy and low-cost polysulfide/iodide redox flow battery. *Nano Energy*, *30*, pp.283–292. https://doi.org/10.1016/j.nanoen.2016.09.043
64. Su, L., Badel, A.F., Cao, C., Hinricher, J.J. and Brushett, F.R., 2017. Toward an inexpensive aqueous polysulfide–polyiodide redox flow battery. *Industrial and Engineering Chemistry Research*, *56*(35), pp.9783–9792. https://doi.org/10.1021/acs.iecr.7b01476
65. Zhu, Y., He, K., Tsega, T.T., Ali, N., Zai, J., Huang, S., Qian, X. and Chen, Z., 2020. Cu_2CoGeS_4 nanocrystals for high performance aqueous polysulfide/iodide redox flow batteries: enhanced selectively towards the electrocatalytic conversion of polysulfides. *Sustainable Energy and Fuels*, *4*(6), pp.2892–2899. https://doi.org/10.1039/C9SE01201J
66. Zai, J., Zhu, Y., He, K., Iqbal, A., Huang, S., Chen, Z. and Qian, X., 2020. Sandwiched Cu_7S_4@ graphite felt electrode for high performance aqueous polysulfide/iodide redox flow batteries: enhanced cycling stability and electrocatalytic dynamics of polysulfides. *Materials Chemistry and Physics*, *250*, p.123143. https://doi.org/10.1016/j.matchemphys.2020.123143
67. Ma, D., Hu, B., Wu, W., Liu, X., Zai, J., Shu, C., Tadesse Tsega, T., Chen, L., Qian, X. and Liu, T.L., 2019. Highly active nanostructured CoS_2/CoS heterojunction electrocatalysts for aqueous polysulfide/iodide redox flow batteries. *Nature Communications*, *10*(1), p.3367. https://doi.org/10.1038/s41467-019-11176-y

68. Bakthavatchalam, B., Habib, K., Saidur, R., Aslfattahi, N. and Rashedi, A., 2020. Investigation of electrical conductivity, optical property, and stability of 2D MXene nanofluid containing ionic liquids. *Applied Sciences*, *10*(24), p.8943. https://doi.org/10.3390/app10248943
69. Pensado, A.S. and Pádua, A.A., 2011. Solvation and stabilization of metallic nanoparticles in ionic liquids. *Angewandte Chemie International Edition*, *50*(37), pp.8683–8687. https://doi.org/10.1002/anie.201103096
70. Stankus, D.P., Lohse, S.E., Hutchison, J.E. and Nason, J.A., 2011. Interactions between natural organic matter and gold nanoparticles stabilized with different organic capping agents. *Environmental Science and Technology*, *45*(8), pp.3238–3244. https://doi.org/10.1021/es102603p
71. Astruc, D., 2008. *Nanoparticles and catalysis* (Vol. 1). Wiley-Vch Verlag, Weinheim. https://doi.org/10.1002/9783527621323
72. Jiang, J., Oberdörster, G. and Biswas, P., 2009. Characterization of size, surface charge, and agglomeration state of nanoparticle dispersions for toxicological studies. *Journal of Nanoparticle Research*, *11*, pp.77–89. https://doi.org/10.1007/s11051-008-9446-4
73. Kim, T., Lee, K., Gong, M.S. and Joo, S.W., 2005. Control of gold nanoparticle aggregates by manipulation of interparticle interaction. *Langmuir*, *21*(21), pp.9524–9528. https://doi.org/10.1021/la0504560
74. He, Z. and Alexandridis, P., 2015. Nanoparticles in ionic liquids: interactions and organization. *Physical Chemistry Chemical Physics*, *17*(28), pp.18238–18261. https://doi.org/10.1039/c5cp01620g
75. Bakthavatchalam, B., Habib, K., Wilfred, C.D., Saidur, R. and Saha, B.B., 2022. Comparative evaluation on the thermal properties and stability of MWCNT nanofluid with conventional surfactants and ionic liquid. *Journal of Thermal Analysis and Calorimetry*, *147*, pp.393–408. https://doi.org/10.1007/s10973-020-10374-x
76. Wegner, S. and Janiak, C., 2017. Metal nanoparticles in ionic liquids. *Topics in Current Chemistry*, *375*, p.65. http://doi.org/10.1007/s41061-017-0148-1
77. Manojkumar, K., Sivaramakrishna, A. and Vijayakrishna, K., 2016. A short review on stable metal nanoparticles using ionic liquids, supported ionic liquids, and poly(ionic liquids). *Journal of Nanoparticle Research*, *18*, p.103. http://doi.org/10.1007/s11051-016-3409-y
78. Kang, S.W., Char, K. and Kang, Y.S., 2008. Novel application of partially positively charged silver nanoparticles for facilitated transport in olefin/paraffin separation membranes. *Chemistry of Materials*, *20*(4), pp.1308–1311. https://doi.org/10.1021/cm071516l
79. Shaikh, S., Lafdi, K. and Ponnappan, R., 2007. Thermal conductivity improvement in carbon nanoparticle doped PAO oil: an experimental study. *Journal of Applied Physics*, *101*(6), p.064302. https://doi.org/10.1063/1.2710337
80. Murshed, S.M.S., de Castro, C.A.N., Lourenço, M.J.V., França, J., Ribeiro, A.P.C., Vieira, S.I.C. and Queirós, C.S., 2011. Ionanofluids as novel fluids for advanced heat transfer applications. *International Scholarly and Scientific Research and Innovation*, *5*(4), pp.579–582.
81. Maxwell, J.C., 1873. *A treatise on electricity and magnetism* (Vol. 1). Clarendon Press, London.
82. Hamilton, R.L. and Crosser, O.K., 1962. Thermal conductivity of heterogeneous two-component systems. *Industrial and Engineering Chemistry Fundamentals*, *1*(3), pp.187–191. https://doi.org/10.1021/i160003a005
83. Bruggeman, V.D., 1935. Berechnung verschiedener physikalischer Konstanten von heterogenen Substanzen. I. Dielektrizitätskonstanten und Leitfähigkeiten der Mischkörper aus isotropen Substanzen. *Annalen der physik*, *416*(7), pp.636–664. https://doi.org/10.1002/andp.19354160705

84. Joseph, A., Sobczak, J., Żyła, G. and Mathew, S., 2022. Ionic liquid and ionanofluid-based redox flow Batteries—A mini review. *Energies*, *15*(13), p.4545. https://doi.org/10.3390/en15134545
85. Lobato, J., Oviedo, J., Cañizares, P., Rodrigo, M.A. and Millán, M., 2020. Impact of carbonaceous particles concentration in a nanofluidic electrolyte for vanadium redox flow batteries. *Carbon*, *156*, pp.287–298. https://doi.org/10.1016/j.carbon.2019.09.045
86. Lee, W., Lee, J., Min, S.K., Park, T., Yi, W. and Han, S.H., 2009. Effect of single-walled carbon nanotube in PbS/TiO_2 quantum dots-sensitized solar cells. *Materials Science and Engineering: B*, *156*(1-3), pp.48–51. https://doi.org/10.1016/j.mseb.2008.11.014
87. Kędra-Królik, K., Cesari, L., Mutelet, F. and Rogalski, M., 2018. Capacity enhancement of ionic liquids-based nanofluid for fuels desulfurization purposes. *Industrial and Engineering Chemistry Research*, *57*(43), pp.14718–14726. https://doi.org/10.1021/acs.iecr.8b02905
88. Dong, C., Yu, Y., Zhang, X., Huang, L., Wu, Y., Li, J. and Liu, Z., 2020. An ionic liquid-modified reduced graphene oxide electrode material with favourable electrochemical properties. *New Journal of Chemistry*, *44*(16), pp.6428–6434. https://doi.org/10.1039/D0NJ00914H

3 Ionanofluid Electrolytes for Electrochemical Storage Applications

Subhratanu Bhattacharya and Bula Dutta

3.1 INTRODUCTION

The growing pursuit of sustainable and dependable energy storage technologies has intensified research into advanced electrochemical systems such as batteries and supercapacitors. Among the numerous components determining their efficiency, the electrolyte remains pivotal, governing ionic transport and mediating essential electrode reactions [1]. Conventional aqueous and organic solvent-based electrolytes, though widely used, often suffer from volatility, flammability, and narrow stability ranges, which reduce the overall durability and safety of energy storage systems. To address these challenges, research has increasingly focused on ionic liquids (ILs)—molten salts that remain liquid near room temperature. These materials combine high ionic conductivity and chemical tunability with exceptional thermal stability and non-flammable behavior, making them an attractive alternative to conventional electrolytes [2–4]. Because their properties can be customized through specific combinations of organic cations and inorganic or organic anions, ILs are frequently described as *designer solvents*. This versatility enables their integration into lithium-ion batteries (LIBs), supercapacitors, and next-generation energy storage devices [5]. However, their high viscosity and limited ion mobility at room temperature pose challenges for practical applications [6, 7]. To overcome these limitations, two-phase or composite electrolytes containing a substantial volume fraction of submicron particles have been developed and shown to deliver higher ionic conductivity than single-phase counterparts [7, 8]. Through favorable ion–surface and interfacial interactions, IL–nanoparticle (NP) hybrids, classified as ionanofluid (INF) electrolytes, enhance electrolyte performance—improving conductivity, electrochemical and thermal stability, and diffusion coefficients—thereby enabling effective operation as liquid electrolytes in electrochemical storage devices.

First conceptualized by Nieto de Castro et al. in 2009 [9], INFs represent an innovative hybrid class of electrolytes formed by two widely explored strategies, viz., (i) dispersing nanostructures (e.g., NPs, nanotubes, or nanofibers, carbon-based materials like graphene, or magnetic nanotubes) directly in ILs and (ii) using IL-grafted NPs to stabilize interfaces and tailor transport. Modifying NP surfaces with ILs can greatly enhance their properties, including catalytic activity, adsorption capacity, thermal resistance, and electrical or electrochemical performance [10]. INF electrolytes combine the intrinsic advantages of ILs—high ionic conductivity, wide electrochemical

 DOI: 10.1201/9781003595250-3

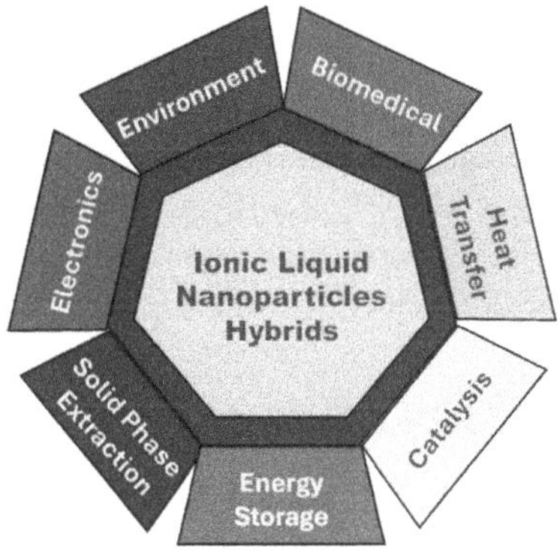

FIGURE 3.1 Key applications of NP-IL hybrids.

stability windows, excellent chemical and thermal stability, negligible vapor pressure, and broad solvation capability—with nanostructure-derived benefits, including mechanical reinforcement, enhanced ion-transport pathways, and augmented electrochemical interfaces [11–14]. These advanced electrolytes have gained traction for their potential to boost energy density in storage devices—reportedly up to 30 times higher than conventional systems—while maintaining high charge–discharge efficiency (>94% over cycles) and operational safety [15]. For instance, dispersions of graphene NPs in vanadium-based IL electrolytes have demonstrated a 14% increase in charge capacity and enhanced power density in redox flow batteries [15]. In electrochemical storage applications, INF electrolytes offer several key advantages over pure ILs or traditional solvents. Their NP inclusions can mitigate high crystallinity [16] and viscosity issues common in ILs [17, 18], thereby improving mass transport and reducing internal resistance, which is critical for high-power devices such as lithium batteries [6, 19] and supercapacitors [20, 21]. In LIBs, INFs enhance lithium intercalation/deintercalation kinetics, leading to better cycling stability and higher cycle stability [6, 19]. For supercapacitors, they enable wider voltage windows and superior capacitance retention at elevated temperatures, addressing limitations of aqueous electrolytes (narrow potential windows) and organic ones (flammability) [21]. These properties make INFs particularly suitable for next-generation applications (Figure 3.1), including electric vehicles, renewable energy integration, and portable electronics, where safety and efficiency are paramount [2].

Despite these benefits, challenges persist in INF development. High costs associated with IL synthesis and NP dispersion, along with potential reductions in electrical conductivity at elevated NP loadings, hinder commercialization. Additionally, issues like agglomeration, viscosity optimization, and long-term stability under operational stresses require further investigation to ensure environmental sustainability and scalability [15].

This chapter aims to provide a comprehensive and timely overview of the electrochemical applications of IL–NP hybrid materials, emphasizing the pivotal role of their structure in governing functional properties. It examines how combining ILs with NPs synergistically enhances their physical, chemical, and electrochemical properties. The discussion begins with an overview of different classes of these hybrid systems, including their design and synthesis methods. Subsequently, the chapter highlights their most extensively explored applications of INFs as electrolytes in electrochemical storage devices. The concluding section critically examines the prospects and challenges of INF electrolytes, outlining strategies to overcome current limitations and offering perspectives on their potential to drive transformative advances in next-generation electrochemical energy storage technologies.

3.2 STRUCTURAL DESIGN OF INFs

A well-dispersed combination of NPs and ILs forms the backbone of INF functionality. Achieving this balance demands control over intermolecular interactions such as hydrogen bonding, electrostatic stabilization, and van der Waals forces. Effective dispersion prevents aggregation, allowing NPs to maintain their high surface area and chemical activity. Stabilization in IL media typically relies on an interplay of electrostatic and steric effects: ions from the IL align around the NPs to offset their surface charges [22]. In these systems, the IL often creates a protective solvation shell around the NPs, enhancing their chemical robustness and thermal stability. In some systems, ILs are chemically grafted onto NPs' surfaces, creating nanoscale hybrid ionic fluids (NHIFs). These hybrid materials retain fluidity while achieving remarkable structural stability and tunable conductivity. The precise design of these IL–NP interfaces—through control of size, charge density, and grafting method—determines how efficiently ions migrate and how stable the electrolyte remains under stress or heat.

The following sections explore the various hybrid structures that can form between ILs and nanomaterials.

3.2.1 NP Dispersions Within ILs

NPs typically display their unique physicochemical properties only when they exist in a well-dispersed state. Due to their high surface energy, NPs tend to diffuse, aggregate, and form bulk-like structures, which suppresses their nanoscale characteristics. Therefore, stabilization strategies are essential to control aggregation and maintain their functional properties. Traditionally, polymers, surfactants, or ionic species have been employed as protective agents. However, the intrinsic self-organization and tunable interactions of ILs provide a favorable alternative, enabling the generation and stabilization of metallic, metal oxide, and semiconducting NPs without the need for additional surfactants.

3.2.1.1 Stabilization Mechanisms in IL Media

Metal and metal oxide NPs have been widely employed as nanoadditives to develop stable IL-based nanofluids, both for research purposes and prospective applications. The stabilization of NPs in ILs arises from multiple, cooperative mechanisms:

3.2.1.1.1 Electrostatic Interactions

NPs can be synthesized and dispersed in ILs with remarkable stability, owing to the inherent ability of ILs to suppress aggregation [23–26]. Since NPs are larger than

IL ions, surrounding cations and anions can organize uniformly at the NP surface, forming electrostatic interactions that counterbalance van der Waals attractions. This electrostatic stabilization effectively suppresses NP aggregation, consistent with the principles of the Derjaguin–Landau–Verwey–Overbeek (DLVO) theory [27].

Imidazolium (Im)-based ILs self-assemble into semiorganized supramolecular aggregates, such as $[(Im)_{x-n}(X)_x]^{n-}$ (anion aggregates, where X denotes the anion) and $[(Im)_x(X)_{x-n}]^{n+}$ (cationic aggregates). These generate electrostatic repulsion consistent with DLVO forces [29]. Beyond these DLVO interactions, the extended alkyl side chains of ILs protrude away from the NP surface, providing an additional steric hindrance. That further prevents NP aggregation [30]. INFs, prepared either by in situ synthesis or by dispersing NPs in ILs, demonstrate significant potential as electrolytes in electrochemical and related energy devices [31]. A notable example is the incorporation of 0.5 wt.% gold NPs into [Emim][$EtSO_4$], which led to a 190% enhancement in electrolytic capacitance along with improved ionic conductivity (Figure 3.2) [28]. These enhancements were attributed to the preferential attraction of $[EtSO_4]^-$ anions to the NP surfaces, thereby modulating the local charge distribution and ionic structure [28]. Similarly, the addition of 0.08 wt.% carbon-encapsulated copper NPs (Cu@C) into [Bmim][PF_6] yielded an INF electrolyte for dye-sensitized solar cells, which exhibited a 35% increase in diffusion coefficient and a 65% rise in electrical conductivity compared to the pristine IL [32]. This was attributed to Cu@C NP interactions with $[Bmim]^+$ cations that enhanced ionic bonding with $[PF_6]^-$ anions, improved cation self-diffusion, and lowered viscosity [32]. Moreover, dispersing silver NPs in imidazolium dicyanamide–based ILs produced INF electrolytes that, when applied in plasmonically enhanced thin dye-sensitized solar cells, exhibited improved stability [33]. In another study, Chen et al. prepared SiC/IL INFs for direct absorption solar collectors and found that thermal properties increased with temperature, while viscosity and density decreased [34]. Beyond the primary interparticle forces, the extended alkyl side chains of the ILs protrude away

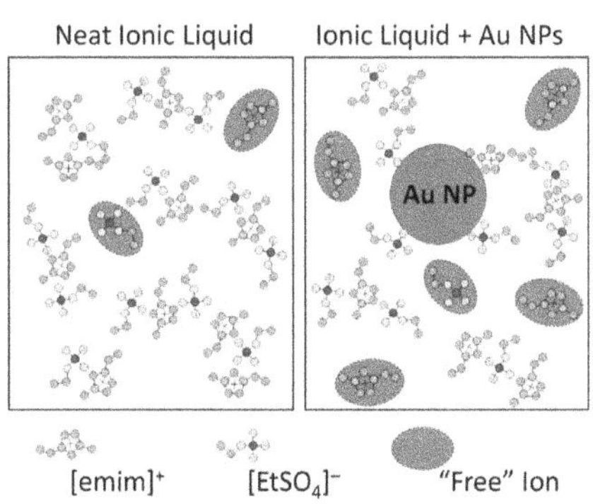

FIGURE 3.2 Schematic illustrating the redistribution of ions following the incorporation of gold NPs into [Emim][$EtSO_4$] IL, where $[EtSO_4]^-$ anions are drawn toward the nanoparticle surface, leading to the liberation of free $[Emim]^+$ cations into the surrounding medium. (Reproduced with permission from [28]. Copyright 2014 Royal Society of Chemistry.)

from the NP surfaces, providing additional steric hindrance that further prevents close particle–particle interactions.

3.2.1.1.2 Hydrogen Bonding and Solvation Forces

Several ILs form highly ordered supramolecular structures through self-association driven by hydrogen bonding. Among them, imidazolium-based ILs exhibit a greater degree of organization compared to quaternary ammonium or phosphonium ILs, owing to their strong tendency to form polymeric aggregates through three-dimensional hydrogen-bonding frameworks. Typically, each cation and anion is surrounded by at least three counterions, as illustrated in Figure 3.3(a) [24]. The strong cohesive energies of oppositely charged moieties promote segregation into polar and non-polar nanodomains. Such preorganized networks can enhance NP solubility by counteracting interparticle attractions. In imidazolium ILs, several sites are available for hydrogen bonding, with the C_2 position being the most acidic [35].

Experimental studies highlight the diversity of hydrogen bonding interactions in ILs. For instance, in $[C_4MIM][PF_6]$, fluorine atoms of the anion participate in (C–H)···F interactions, whereas in $[C_4MIM][CF_3SO_3]$, oxygen atoms dominate (C–H)···O bonding. In $[C_4MIM]$[TFSI], both fluorine and oxygen atoms contribute to hydrogen bonding with the cation. These intrinsic cation–anion interactions, however, compete with the hydrogen bonding established between ILs and NP surfaces, particularly when hydroxy or oxy functionalities are present. Such interactions generate alternating solvation layers of cations and anions around NPs, providing sufficient repulsive forces to prevent aggregation. These distance-dependent solvation forces are visible in both protic and aprotic ILs [35, 37, 38]. Gao et al. demonstrated the stabilization of SiO_2 NPs in $[C_4MIM][BF_4]$ by incorporating fluorocarbon groups (1H,1H,9H-hexadecafluoro-1-nonanol). Small-Angle Neutron Scattering (SANS) and Dynamic Light Scattering, (DLS) confirmed the formation of a seven-layered solvation shell

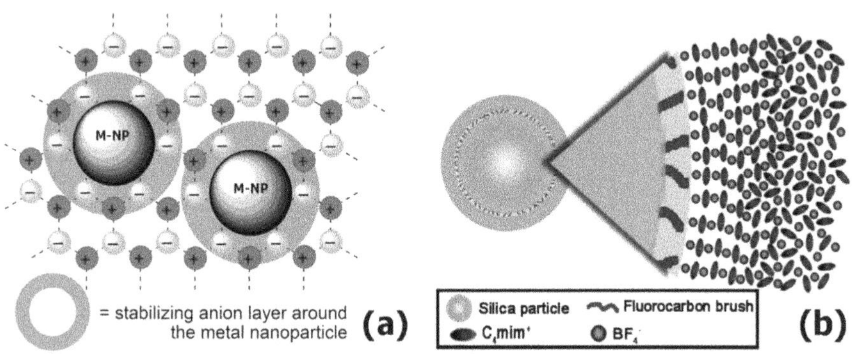

FIGURE 3.3 (a) Stabilization of metal NPs (M-NPs) into the supramolecular IL network by combined electrostatic and steric (electrosteric) interactions, as illustrated by the formation of a primary anion layer surrounding the M-NPs and (b) schematic showing the stabilization of SiO_2 NPs via solvation shell formation in $[C_4MIM][BF_4]$. (Source: (a) Reproduced by permission [24]. Copyright (2011) Elsevier B. V. and (b) reproduced by permission [36]. Copyright (2015) American Chemical Society.)

(≈0.7–0.8 nm per layer; total ≈5 nm), mediated by hydrogen bonding between the fluorocarbon moieties and BF_4^- anions (Figure 3.3(b)) [36]. This multilayered solvation shell effectively imparted steric stabilization. Additional evidence, such as adsorption of $[C_2MIM]^+$ onto TiO_2 at pH < 3.5 through hydrogen bonding between the C_2 position of the imidazolium ring and surface oxygen atoms of rutile TiO_2 [39] further establish it. Likewise, C_2–H···O interactions between imidazolium cations and ZnO surfaces have been implicated in the preferential stabilization and growth of 1D nanostructures such as NPs, nanowires, and nanorods [40].

Apart from hydrogen bonding, solvophobic and structural forces play essential roles in NP dispersion within ILs. Analogous to hydrophobic effects in aqueous systems, solvophobic interactions describe the limited miscibility of surfactants within ILs [41]. Wittmar et al. demonstrated that hydrophobic functionalization of NPs enhances dispersibility specifically in hydrophobic ILs. There, instead of hydrogen bonding, reduced van der Waals attractions favor stable suspensions of alkyl-substituted SiO_2 NPs [42].

Overall, the colloidal stability of NPs in ILs arises from the combined contributions of hydrogen bonding, steric hindrance, solvophobic interactions, van der Waals forces, and electrostatic effects. The balance of these interactions determines the structural organization and dispersion behavior of NPs, giving rise to diverse and tunable IL–NP hybrid architectures.

3.2.1.1.3 NP-IL Interface

The electric double layers (EDLs) at the NP surfaces, where ions accumulate and create charged zones that boost overall conductivity through increased electrophoretic motion and reduced resistance (e.g., up to 70% reduction in electrolyte resistance with 0.5 wt.% gold NPs in $[Emim][EtSO_4]$) [28]. This interface is nanofluidic in nature because the fluid layer around each NP (typically 1–100 nm) experiences confinement effects, leading to anomalous transport like faster ion diffusion or selective adsorption. The EDL thickness, often comparable to the Debye length (1–10 nm in ILs), dominates charge transfer and stability, preventing aggregation via electrostatic repulsion [28, 31]. In addition to boosting electrochemical performance, the formation of EDLs within ILs enhances the overall chemical stability of the electrolytes. These structured ionic layers act as protective barriers at the NP–electrolyte interface, mitigating direct contact with reactive species and thereby preventing NP corrosion.

Figure 3.4 schematically illustrates the formation of ion clusters around NPs, leading to the development of a protective EDL. In this depiction, electrostatic interactions drive the attraction of ions and their associated clusters toward the NP surface. This process results in the formation of a semi-ordered ionic shell that serves as a stabilizing interfacial layer, thereby enhancing the overall stability and performance of the electrolyte in battery systems.

3.2.1.1.4 Colloidal Gel and Colloidal Glass

The high ionic strength of ILs or elevated volume fractions can lead to specific interactions that challenge the stability of NP dispersions, often resulting in the formation of solid-like soft materials. Ueno et al. identified two types of short-range ordered composite materials under specific conditions: colloidal gels and colloidal glasses [44–46]. Colloidal gels form when colloidally unstable particles, dominated by attractive forces,

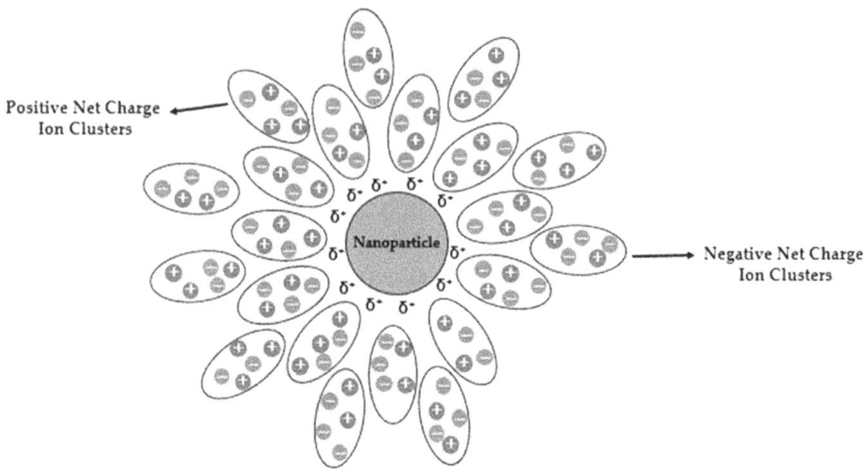

FIGURE 3.4 Schematic illustration of ion clusters assembling around NPs dispersed in ILs, forming a stabilizing electric double layer at the interface. (Source: Reproduced by permission [43]. Copyright 2024 Springer Nature (Open Access).)

aggregate into a fractal-like network that spans the entire suspension volume. In contrast, colloidal glasses arise at high concentrations where colloidally stable particles, governed by repulsive forces, become topologically trapped in cages formed by neighboring particles. Beyond physical adsorption, ILs can chemically bind to NP surfaces through interactions between IL components and functional groups on the NP sorbent surface [44].

3.2.2 IL Grafted NPs: Nanoscale Hybrid Ionic Fluids (NHIFs)

Beyond their role as dispersion media or through simple physical adsorption on NPs, ILs can also be chemically grafted onto NP surfaces. When IL layers are immobilized on solid supports such as NPs, the resulting systems are known as NHIFs [47–54]. NHIFs are functionalized hybrid nanostructures that exhibit solvent-free, liquid-like behavior while retaining their nanoscale organization in the fluid state. In this distinctive class of materials, the intrinsic surface charge of the NP core can directly anchor the IL (canopy), resulting in a soft-gel architecture. This arrangement not only reduces NP aggregation by providing a steric barrier but also enables the IL canopy to impart fluidity, stability, and tailored functionality to the overall system. By modulating the canopy's grafting density, a solid-to-liquid transition can be achieved, resulting in a macroscopically homogeneous single-component fluid. Importantly, the ionic interactions between core and canopy endow NHIFs with dynamic, liquid-like characteristics closely resembling those of room-temperature ILs.

For example, when carbon nanotube bundles are mechanically mixed with ILs, they disentangle, forming finer bundles that create a networked gel structure [55]. This phenomenon likely stems from specific interactions between the imidazolium ions in the ILs and the π-electronic surface of the nanotubes.

The tunability of NHIFs offers an enormous design space for creating multifunctional materials. Key variables include the following:

- *Core characteristics*: composition (metals, ceramic oxides, sulfides, chalcogenides, carbon nanostructures, etc.), morphology, and size.
- *Canopy functionality*: chemical nature and molecular weight.

In the grafting method, a reactive group (X), terminated with a charged functional moiety (Z), directly self-assembles onto the NP core surface. This technique is more straightforward, requiring only one step for functionalization; however, achieving a pure NHIF is often tricky. Issues may arise due to incompatibility between the core and ligand, unwanted side reactions of the terminal group (Z) with the core surface, or steric hindrance effects, which restrict grafting density. Despite these challenges, the availability of diverse commercial coupling agents (such as silanes, phosphates, and chelating moieties) has made this approach more accessible [47] (Figure 3.5).

For example, Yan et al. functionalized SiO_2 NPs with four different ILs using a grafting approach [47]. Two distinct reaction schemes were employed. In the first scheme (Figure 3.5(a)), the coupling agent (3-chloropropyl) trimethoxysilane was reacted with a methyl imidazolium cation (MIM^+) to form a functionalized ligand. This ligand was then covalently anchored to the SiO_2 surface, yielding SiO_2–MIM–Cl. Subsequent anion-exchange reactions produced two NHIFs: SiO_2–IL–NTf_2 (1) and SiO_2–IL–PF_6 (2).

In the reverse scheme (Figure 3.5(b)), (3-chloropropyl) trimethoxysilane first reacted with the trifluoromethanesulfonyl imide (NTf_2^-) anion in the presence of LiH, generating an anion-functionalized, Li^+-neutralized ligand. After covalent grafting to the SiO_2 surface, the intermediate SiO_2–NTf–Li underwent cation-exchange reactions to yield NHIFs containing long-chain cations, specifically SiO_2–IL–BMIM (3) and SiO_2–IL–NBu_4 (4).

Similar to Yan's strategy, Linden Archer and coworkers grafted 1-trimethoxysilylpropyl-3-methylimidazolium bis(trifluoromethylsulfonyl)imide (SpmIm–TSFI) onto SiO_2 NPs [49, 50], achieving a notably higher grafting density (0.8 ligands nm^{-2}). In another study, ZrO_2 NPs were employed as cores, onto which 1-undecyltrimethoxysilane-3-butylimidazolium bis(trifluoromethylsulfonyl)imide ILs were tethered [48]. Bhattacharya et al. grafted 1-ethyl-3-(3-trimethoxysilylpropyl) imidazolium bis(trifluoromethanesulfonyl) imide IL onto two different NP cores, SiO_2 and Al_2O_3 [53]. The grafting density was markedly higher for SiO_2 (≈2.31 chains nm^{-2}) than for Al_2O_3 (≈0.97 chains nm^{-2}), a difference attributed to variations in the intrinsic surface charge of the NPs. The dispersion states of NPs in the NHIFs were further examined using transmission electron microscopy (TEM). As shown in Figure 3.6(a) and (b), the adopted tethering strategy proved effective in anchoring the core NPs with minimal aggregation, thereby yielding gel-like, highly viscous fluids (insets). Notably, the SiO_2-based NHIF (Figure 3.6(a)) exhibited a more uniform, non-aggregated dispersion than its Al_2O_3 analogue (Figure 3.6(b)). This enhanced stability arises from the relatively stronger and chemically robust interactions between the highly curved SiO_2 surfaces and the silane termini of the IL

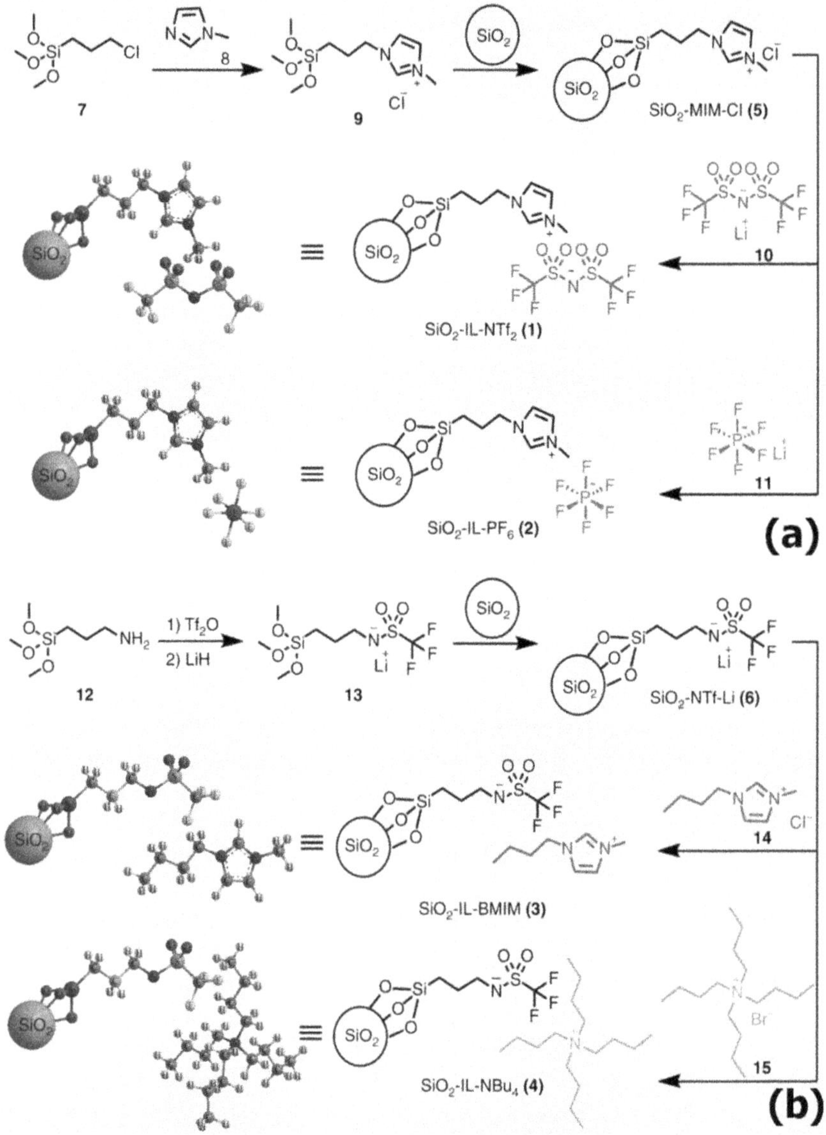

FIGURE 3.5 Different grafting approaches adopted for the synthesis of SiO_2-NHIFs. (a) Through the synthesis of cation-functionalized, anion-neutralized ligand, and (b) through the synthesis of anion-functionalized, cation-neutralized ligand. (Source: Reproduced with permission [47]. Copyright (2017) Springer Nature (Open access).)

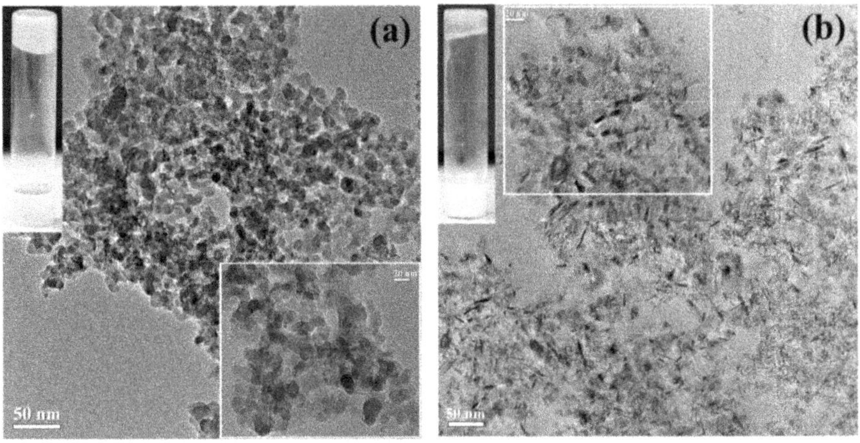

FIGURE 3.6 TEM images of (a) SiO_2–NHIF and (b) Al_2O_3–NHIF. Enlarged regions of the micrographs along with corresponding photographs of the NHIFs are shown in the insets. (Source: Reproduced with permission [53]. Copyright (2017) Elsevier B. V.)

chains, compared to the comparatively weaker tethering observed at Al_2O_3 surfaces. Further investigations extended this methodology to other oxides, such as SnO_2 [52], ZrO_2 [51], ZnS [17], and TiO_2 [54], where different IL chains were similarly grafted to the NP surfaces.

Building upon this approach, hybrid electrolytes incorporating 1-methyl-3-propylimidazolium (IM) and 1-methyl-3-propylpiperidinium (PP) were synthesized to further optimize electrochemical functionality (Figure 3.7). These systems were systematically evaluated in terms of their ionic conductivity, electrochemical stability, mechanical integrity, and capacity to enable uniform and stable lithium electrodeposition [56]. The detailed evaluation of their performance in lithium-based batteries is discussed in subsequent sections.

By systematically tailoring different parameters, researchers have already developed a wide variety of NHIF-based materials with controllable structures and properties, highlighting their vast potential in advanced applications such as sensors, capacitors, actuators, and electrochemical storage [48, 50–58].

3.2.3 INFs Formed by Dispersion of NHIFs in ILs

Another class of very stable INFs can be developed by dispersing different NPs, densely functionalized with silyl-terminated ILs (NHIFs) in different IL hosts. This class of INF was first reported by Archer and his group [57] and later extensively expanded by Bhattacharya and his group [16–18]. A key observation was that the incorporation of even trace amounts of NHIF effectively suppressed crystallization of the host IL. The study of cold crystallization dynamics of NHIF incorporated INF systems revealed the existence of a diffusion-controlled nucleation

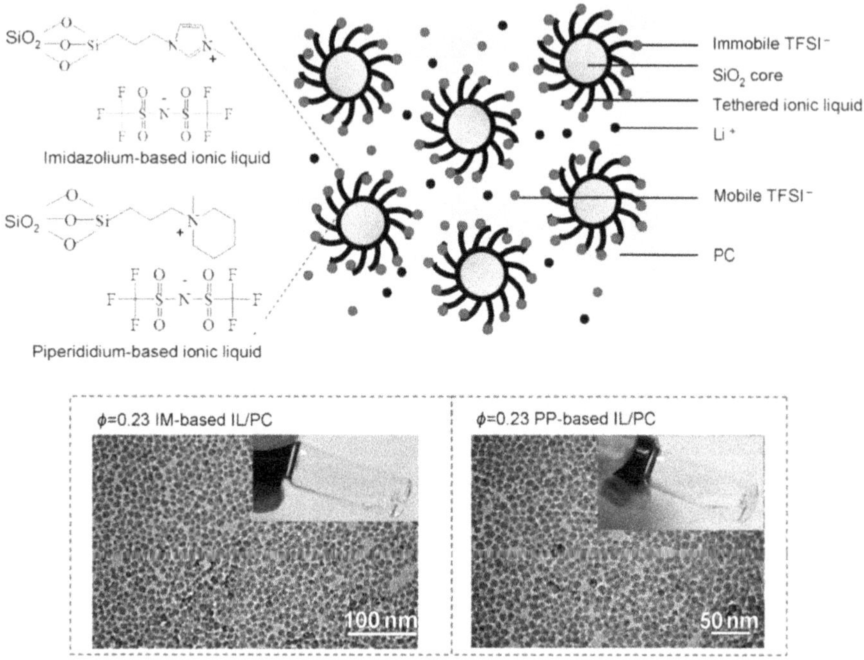

FIGURE 3.7 A schematic illustration of the SiO_2–IL–TFSI/propylene carbonate (PC) hybrid electrolyte system containing 1 M LiTFSI. TEM images corresponding to electrolyte samples with a SiO_2 NP volume fraction of 0.23 are presented: SiO_2–IM–TFSI (bottom left) and SiO_2–PP–TFSI (bottom right). The insets depict photographs of the two hybrid electrolyte formulations, highlighting their homogeneous, fluid-like characteristics, and visually uniform liquid morphology. (Source: Reproduced with permission [56]. Copyright (2014) Wiley-VCH.)

process, which delayed nucleation of IL crystals to higher temperatures, and the presence of higher NHIF content induces a transformation from a rigid, glassy supercooled phase into a flexible, rubbery amorphous liquid phase, circumventing the crystallization process [16]. The resulting hybrid fluids retained liquid-like transport properties over a significantly extended temperature range, thereby enabling the design of ILs with broader operational windows. Using Differential Scanning Calorimetry (DSC), viscosity measurements, and relaxation dynamics, the influence of uniformly dispersed NHIFs on the thermophysical and electrical transport properties of INFs was analyzed [17, 18] (Figure 3.8).

The analysis reveals that, unlike the structural dynamics–governed ionic transport typically observed in pure ILs, NHIF incorporated INFs exhibit a strong decoupling of conductivity relaxation, ion diffusion, and overall conductivity from structural relaxation or viscosity. This decoupling effect becomes more pronounced as the glass transition temperature and fragility index of the INFs increase. The reduced extent of ion-pair formation provides a qualitative explanation for the origin of this behavior [17, 18].

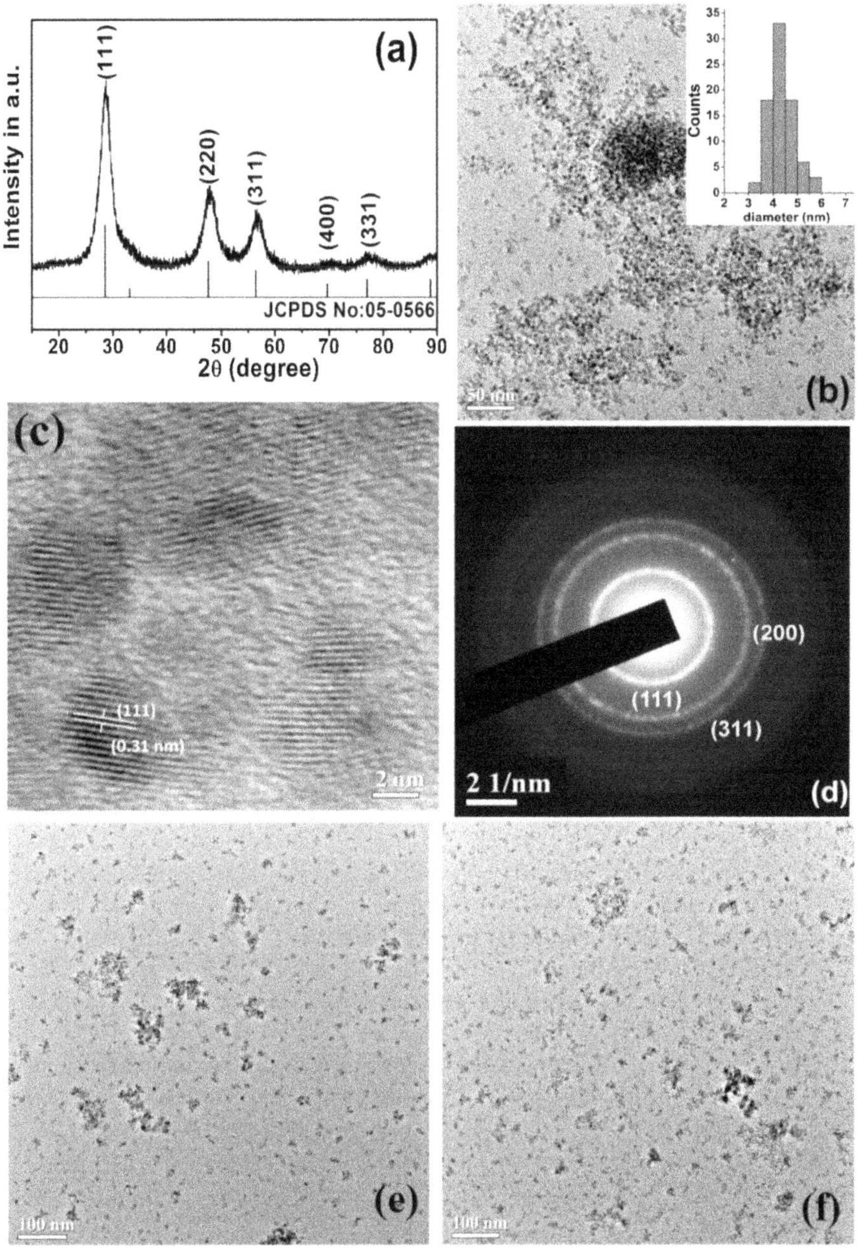

FIGURE 3.8 (a) XRD profile of as-prepared ZnS NPs, (b) TEM images of the ZnS-NHIF with size distribution (inset), (c and d) high-resolution image of ZnS-NHIF and corresponding SAED pattern, respectively, and (e and f) TEM images of 5.0 ZnS-IL and 15.0 ZnS-IL nanofluids, respectively. (Reproduced with permission from [17]. Copyright (2019) Elsevier B. V.)

3.3 APPLICATIONS OF IL-NPS HYBRID ELECTROLYTES IN ELECTROCHEMICAL STORAGE DEVICES

IL-NPs hybrid or INF electrolytes containing a substantial concentration of NPs as an insoluble second phase demonstrate significantly enhanced ionic conductivity compared to conventional IL electrolytes. The fundamentals of these hybrid electrolytes lie in the interfacial interactions between the IL ions and NPs, which create structured nano-regions or "nanofluidic interfaces." These interfaces facilitate ion transport, improve cation diffusion coefficients, boost electrical conductivity [5], mitigate aggregation, enhance electrochemical and thermal stability [59], and provide higher diffusion coefficients, addressing limitations of traditional electrolytes like volatility in organic solvents or narrow potential windows in aqueous systems [3]. Such improvements make these hybrid fluids highly suitable for various electrochemical applications, such as batteries and supercapacitors.

3.3.1 Applications in Lithium Batteries

An ideal electrolyte solvent for a lithium battery must satisfy a combination of physicochemical, electrochemical, and safety conditions to ensure efficient ion transport, stability, and longevity. It must combine high ion transport ability, broad electrochemical stability, and excellent safety characteristics, while being chemically compatible with all cell components.

Key Conditions for an Ideal Electrolyte Solvent

- The solvent must promote high lithium-ion mobility over a wide temperature range, minimizing internal resistance and polarization during charging and discharging.
- Must have a high dielectric constant and low viscosity. A high dielectric constant enhances lithium salt dissociation, while low viscosity facilitates rapid ionic transport, ensuring better conductivity.
- The solvent must be stable within 0–5 V versus Li/Li^+ to prevent side reactions and decomposition at both electrodes during high-voltage operation.
- It should withstand high operating temperatures without decomposition and remain chemically inert toward electrodes, separators, binders, and current collectors.
- The solvent must exhibit oxidation resistance at the cathode and reduction stability at the anode to reduce unwanted reactions, corrosion, or gas evolution.
- The electrolyte should enable the controlled formation of a solid electrolyte interphase (SEI) on the anode, which is ionically conductive but electronically insulating, protecting against further degradation.
- To enhance safety, solvents should have low vapor pressure, high flash points, or ideally be non-flammable to prevent thermal runaway.
- The ideal solvent should be non-toxic, environmentally benign, and cost-effective for large-scale battery production.

ILs have gained attention as potential replacements for conventional organic electrolyte solvents in lithium battery electrolytes because the weak interactions between their bulky cations and charge-delocalized anions give rise to distinctive properties such as non-flammability, chemical inertness, broad electrochemical stability, and high ionic conductivity even at moderate temperatures [60–62]. Despite these attractive features, the widespread practical use of ILs remains challenging due to certain drawbacks. In particular, they exhibit poor mechanical strength and low lithium-ion transference numbers, which limit their efficiency in electrochemical systems. This limitation is primarily linked to the binding of IL components—such as the ether oxygen atoms in Bis(trifluoromethanesulfonyl)imide (TFSI)/Lithium bis(trifluoromethanesulfonyl)imide (LiTFSI) systems—with Li^+ ions. As a result, the low lithium-ion transference number ($t_{Li}{}^+$) indicates that only a small fraction of charge transport is carried by Li^+ ions, thereby reducing the overall ionic mobility and lowering the performance of the battery.

In addition, the broad liquidus range of ILs restricts their practical utility, primarily due to challenges such as electrolyte leakage, complex packaging requirements, and limited portability. During battery operation, ILs can also induce undesirable effects, including lithium dendrite formation, uneven lithium re-deposition, and diminished power output. To address these issues, recent research efforts have focused on mitigating such limitations by grafting or tethering IL molecules onto NP surfaces through electrostatic interactions, thereby enhancing structural stability and electrochemical performance.

Researchers have established that lithium dendrites tend to form within regions of charge imbalance, arising from local anion depletion in the electrolyte during cycling. To suppress such uneven lithium deposition, it is essential to confine anion mobility and maintain a uniform spatial distribution throughout cell polarization. Considerable progress has been made toward designing electrolytes in which anions are anchored to NP surfaces, thereby restricting their motion and rendering lithium ions the primary charge carriers. These systems act as dynamic reservoirs, capable of releasing anions when needed to neutralize space charge accumulation during polarization [63–65].

A variety of hybrid IL–NP electrolytes have been developed by grafting IL molecules onto metal oxide supports such as TiO_2, Al_2O_3, ZrO_2, SiO_2, and Fe_2O_3. This approach combines the electrochemical advantages of ILs with the structural robustness of NPs, overcoming the individual shortcomings of each component [48, 56–58]. The immobilized counterions within these composites serve as stable anionic reservoirs, enabling fast replenishment during charge–discharge processes. Interaction between NPs and both lithium salts and IL anions enhances salt dissociation, resulting in improved rate performance and cycling stability.

Furthermore, the reduced ion-pairing between Li^+ and immobilized bulky anions enhances lithium-ion dissociation and establishes efficient ionic transport pathways. For instance, Lu et al. produced a jammed hybrid electrolyte (SiO_2–IL–TFSI) offering a stable anion reservoir that effectively suppressed dendrite growth [50]. Remarkably, its ionic conductivity was approximately three times higher than that of the individual IL or NP components. Similarly, Delacroix et al. increased the lithium-ion transference number ($t_{Li}{}^+$) to 0.56 by anchoring imidazolium-based

ILs onto SiO_2 NPs, achieving conductivities of 10^{-6} S cm^{-1} at low temperature and 10^{-4} S cm^{-1} at intermediate temperature [66]. Further research by Lu et al. on IL–NP hybrids incorporating [C_3MIM][TFSI] and [MPP][TFSI] demonstrated that piperidinium-based systems exhibited superior conductivity, enhanced cycling life, and effective dendrite suppression during prolonged operation. In contrast, imidazolium-based formulations enabled dendrite-free lithium electrodeposition but required a higher IL or NP content [56, 67]. The brush-like architecture of ILs tethered to NPs has been shown to significantly enhance the dispersion of silica NPs within conventional propylene carbonate (PC)/LiTFSI electrolytes [67]. This improvement is primarily attributed to the combined effects of steric hindrance and electrostatic double-layer interactions, which stabilize the NP suspension and promote uniform distribution throughout the electrolyte matrix. The introduction of these nanometer-scale, impermeable particles results in the formation of a mechanically reinforced and highly tortuous porous network, capable of effectively impeding lithium dendrite growth. Such structural reinforcement plays a crucial role in preventing dendrite penetration that could otherwise bridge the electrodes, leading to short circuits and potential thermal runaway [67]. Similarly, the incorporation of 10% 1-methyl-3-propylimidazolium bis(trifluoromethanesulfonyl)imide (IL)-tethered SiO_2 NPs into a PC–NaTFSI electrolyte markedly improved the cycling performance of Na–CO_2/O_2 batteries [68]. The system achieved over 20 stable charge–discharge cycles at voltages up to 5 V without observable electrolyte degradation, demonstrating the beneficial synergism between IL functionality and NP reinforcement [68]. Further refinements have focused on covalently grafting long-chain imidazolium-based ILs onto metal oxide cores such as zirconia (ZrO_2, 86 ± 2 nm). These hybrid materials exhibit enhanced electrochemical and thermal stability while simultaneously reinforcing the mechanical integrity of the electrolyte. The embedded ZrO_2 NPs increase rigidity, thereby resisting dendrite propagation and improving interfacial stability at the electrode–electrolyte boundary [48].

Electrochemical performance analysis of the IL–NP hybrid electrolytes containing IM and PP cations (Figure 3.7) revealed distinct differences in their behaviors as shown in Figure 3.9. The PP-based hybrid electrolytes exhibited higher ionic conductivity and demonstrated a superior ability to suppress cell short-circuiting compared to their IM-based counterparts. Incorporation of as little as 11 wt.% of SiO_2–PP–TFSI into a PC–LiTFSI host yielded more than a tenfold enhancement in cell lifetime, underscoring the efficiency of the piperidinium framework.

Meanwhile, IM-based electrolytes displayed comparable suppression of lithium dendrite formation but required higher concentrations of IL and/or NPs to achieve similar results (Figure 3.10). Remarkably, both electrolyte systems provided enhancements in cell lifetime exceeding 10,000-fold beyond what could be predicted from their mechanical modulus alone. Furthermore, galvanostatic cycling of Li/$Li_4Ti_5O_{12}$ half cells using these IL–NP hybrid electrolytes demonstrated over 500 stable cycles with improved rate capability and sustained electrochemical stability, highlighting their potential for next-generation lithium metal batteries.

Bhattacharya and his coworkers developed an innovative class of INF electrolytes for safer, high-energy-density lithium metal batteries (LMBs). These electrolytes were engineered by dispersing TiO_2 [6] and SnO_2 [19] tethered NHIFs into

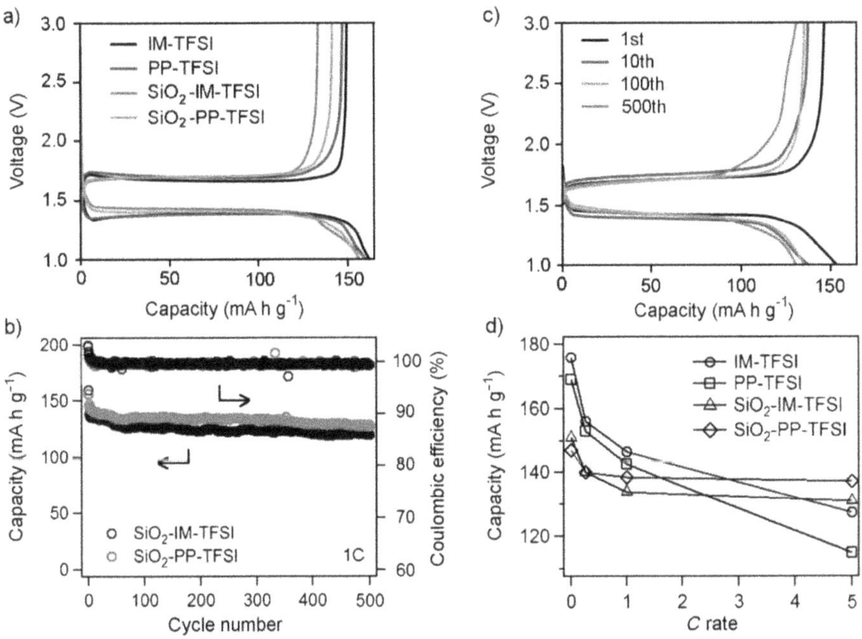

FIGURE 3.9 Charge–discharge behavior of Li/LTO full cells employing IM–TFSI, PP–TFSI, SiO_2–IM–TFSI, and SiO_2–PP–TFSI electrolytes at a volume fraction of 0.23. (a) The initial charge–discharge profiles were recorded at a current density of 0.315 mA cm^{-2} (1C rate), (b) the variation of discharge capacity with cycling number is presented for both hybrid electrolyte systems, demonstrating their stability over repeated charge–discharge cycles, (c) representative charge–discharge curves for the initial, 10th, 100th, and 500th cycles of SiO_2–PP–TFSI electrolytes in Li/LTO cells at room temperature highlight the retention of electrochemical performance, and (d) the dependence of cell capacity on charge rate (C-rate) for all four electrolyte systems is also depicted, underscoring their rate-capability characteristics. (Source: Reproduced with permission [56]. Copyright (2014) Wiley-VCH.)

high-salt-concentration IL matrices based on $[Pyr_{14}]$[TFSI] and [DMim][TFSI]. Beyond the intrinsic advantages of ILs—such as thermal stability, non-flammability, and wide electrochemical windows—the resulting INF systems exhibited several distinctive features valuable for LMB performance. Among these were viscosity-decoupled ionic conductivity and suppression of salt-induced crystallinity, features that enable efficient lithium-ion transport even at elevated salt concentrations.

The INF electrolytes demonstrated superior ionic conductivity to pure IL electrolytes while requiring significantly lower lithium salt content. Moreover, they exhibited an enhanced lithium-ion transference number (t_{Li}^+) under ambient conditions, promoting balanced charge migration within the electrolyte. Electrochemical studies revealed that a Li/$LiMn_2O_4$ cell containing the optimized $[Pyr_{14}]$[TFSI]-based electrolyte retained over 96% of its theoretical discharge capacity after 50 cycles at a current density of 24 mA g^{-1}, outperforming a conventional 0.2 M salt-dissolved IL system, which retained only 88% capacity (Figure 3.11). Similarly, a Li/$LiFePO_4$ cell employing [DMim][TFSI]-based electrolytes achieved over 96% capacity retention

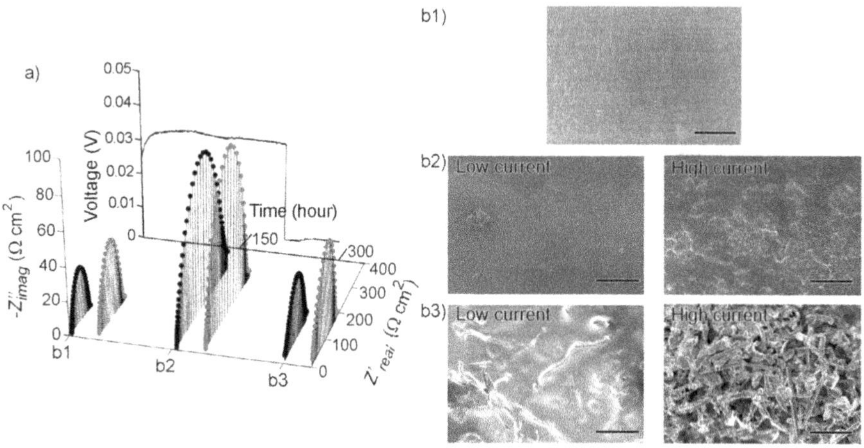

FIGURE 3.10 Potential profiles recorded at a fixed current density, along with impedance spectra and SEM micrographs, are presented for IM–TFSI and SiO_2–IM–TFSI. Data correspond to three distinct stages: (a) Potential profiles recorded at a fixed current density, along with impedance spectra and (b) SEM micrographs (b1, b2 and b3), are presented for IM–TFSI and SiO_2–IM–TFSI. Data in Figure 3.10 a) correspond to three distinct stages, viz., prior to polarization-b1, during steady-state operation-b2, and following cell short-circuit-b3. The SEM images corresponding to b1, b2 and b3 stages on the right display electrode surface morphologies obtained after low-current density (0.02 mA cm^{-2}) and high-current density (0.1 mA cm^{-2}) cycling. All SEM images were captured at identical magnification, with a scale bar of 100 μm. (Source: Reproduced with permission [56]. Copyright (2014) Wiley-VCH.)

with a coulombic efficiency of 99% at a C/5 rate after 50 cycles, substantially exceeding the 81% observed with its pure IL counterpart (Figure 3.12).

Overall, the integration of IL-tethered NPs represents a strategically engineered approach to address the long-standing trade-offs between ionic conductivity, mechanical robustness, and dendrite suppression in liquid and quasi-solid electrolytes. Such hybrid systems illustrate a promising pathway toward the development of next-generation high-performance and safe lithium-based batteries.

3.3.2 Applications in Supercapacitors

Supercapacitors have emerged as promising alternatives to conventional rechargeable energy storage devices, particularly in renewable energy generation and storage systems, owing to their ultra-high-power density, rapid charge–discharge capability, and exceptional cycle life. Despite these advantages, the low energy density of supercapacitors remains a major limitation that restricts their widespread application across large-scale energy systems. Enhancing their performance therefore necessitates the development of electrode materials with high specific capacitance and excellent rate capability, alongside the design of electrolytes offering wide electrochemical stability windows. Such advancements are vital to bridge the performance gap between batteries and supercapacitors, ensuring both high power delivery and adequate energy storage capacity for integration in next-generation renewable energy grids, hybrid storage systems, and electric mobility applications.

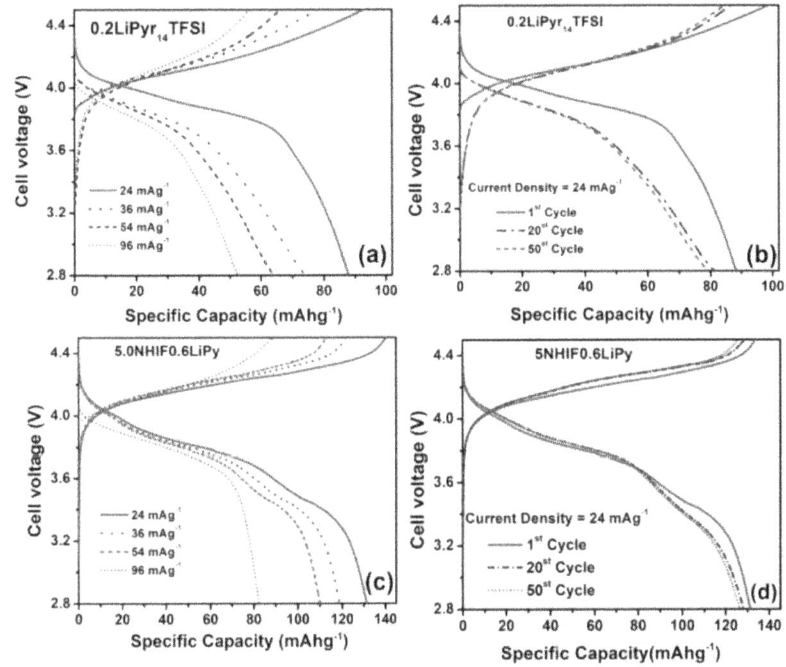

FIGURE 3.11 Charge–discharge characteristics of $Li/LiMn_2O_4$ cells employing (a and b) $0.2LiPyr_{14}TFSI$ and (c and d) 5NHIF0.6LiPyr electrolytes at 25°C, are presented at various current densities and selected cycling stages. The profiles illustrate the evolution of electrochemical performance across different charge–discharge cycles, highlighting the superior rate capability and sustained discharge behavior of the ionanofluid-based electrolyte (5NHIF0.6LiPyr) compared to the conventional IL-based system ($0.2LiPyr_{14}TFSI$). (Reproduced with permission from [6]. Copyright (2018) Elsevier B. V.)

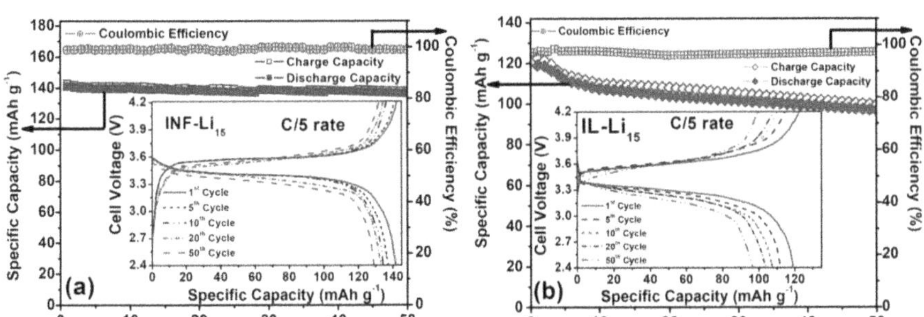

FIGURE 3.12 Voltage profiles of (a) $Li/INF–Li_{15}/LiFePO_4$ and (b) $Li/INF–Li_{10}/LiFePO_4$ cells recorded at various current rates, along with their corresponding rate capability plots. The comparative profiles demonstrate the evolution of discharge potential and capacity with increasing current densities, emphasizing the enhanced rate performance and electrochemical stability of the INF-based electrolytes under dynamic cycling conditions. (Reproduced with permission from [19]. Copyright (2023) Elsevier B. V.)

The energy storage mechanism in supercapacitors primarily depends on ion dynamics and physicochemical interactions at the electrolyte–electrode interface. The incorporation of IL crystal electrolytes (ILCEs) has been shown to lower charge transfer resistance and increase both the charge storage and discharge duration of the devices, thereby enhancing their overall energy storage efficiency. In IL-based systems, double-layer formation occurs through physical capacitive electrosorption between charged ions and the inert electrode surface, eliminating the need for additional solvent molecules to facilitate charge transport.

The theoretical operational voltage of an electrochemical double-layer capacitor (EDLC) is governed by the upper potential limit (P_U) and lower potential limit (P_L), beyond which oxidation or reduction of the electrolyte may occur. Ideally, the maximum operating voltage corresponds to the potential difference between P_U and P_L (i.e., $P_U - P_L$). However, in practical operation, the full theoretical voltage range cannot be fully utilized due to electrochemical instability and asymmetric electrode behavior [69–71].

In an EDLC, the potential of zero voltage (P_0V)—defined as the potential of both electrodes when the device operates at 0 V—does not necessarily align with the midpoint of the P_U–P_L range. As a result, during charging, one electrode often reaches its potential limit (P_U or P_L) earlier, while the other remains within an unexploited potential region. This imbalance causes the actual operational voltage to fall below the theoretical value, which in turn leads to a reduction in energy density [71].

Furthermore, operating an EDLC at its full theoretical voltage typically accelerates electrolyte degradation and electrode instability, resulting in poor cycling performance and device failure. Therefore, optimizing the potential window and balancing the potential ranges of the two electrodes are critical strategies to maximize the usable voltage and improve both the energy density and long-term stability of EDLCs.

The incorporation of IL–NP hybrids as electrolytes or electrolyte additives has proven to be one of the most effective strategies for optimizing the potential ranges of EDLC electrodes. Such hybrid systems enable synergistic interactions between the IL ions and NP surfaces, thereby stabilizing electrode potentials, suppressing electrolyte decomposition, and extending the accessible electrochemical window. This dual functionality enhances both the energy density and cycling stability of the device, representing a promising route for the development of high-voltage, high-performance supercapacitors [51–54, 71].

Bhattacharya and his group investigated the mechanical, electrochemical, and thermal stabilities of three NHIFs obtained by covalent attachment of imidazolium or pyrrolidinium cation-based ILs on different nanocores at different grafting densities [51–54]. The temperature-dependent electrochemical properties of the NHIF electrolytic cells have been studied within planar steel electrodes [51–53] and in graphite electrodes [54]. A wide electrochemical stability window (8 V) has been observed in most of the NHIFs, which do not shrink even at higher temperatures. The value of the specific capacitance of the cells is found to be of the order 10^{-5} F cm^{-2}. Both the specific capacitance and conductivity show a thermally activated nature, presumably due to increasing ion mobility and decreasing electrode–electrolyte interfacial resistance with increasing temperature. Moreover, the electrolytic cells show long-term cycling stability after 3000 charge/discharge cycles at 50°C with 90% and 80.5% capacitance retention. Thus, the interfacial mobility of charge carriers within both the thermally stable and mechanically compliant NHIFs is reasonably sufficient for their application as electrolytes for high-power-density secondary charge storage devices, even at higher temperatures. Finally, room temperature electrochemical performance of symmetric Carbonaceous Mesophase Spherule (CMS) graphite electrode supercapacitors [54] based on this innovative hybrid electrolyte is found to be comparable or better than that of the IL-based supercapacitors reported in the literature (Figure 3.13).

In a similar study [71], Dou et al. developed symmetric activated carbon (AC)-based EDLC, where the potential of zero voltage (P_0V) was effectively tuned by incorporating SiO_2–IL–NTf_2 into a BMIM–NTf_2/PC electrolyte. This adjustment enabled the potential ranges of the two AC electrodes to be optimized within the electrochemical stability window. Consequently, the modified EDLC could operate

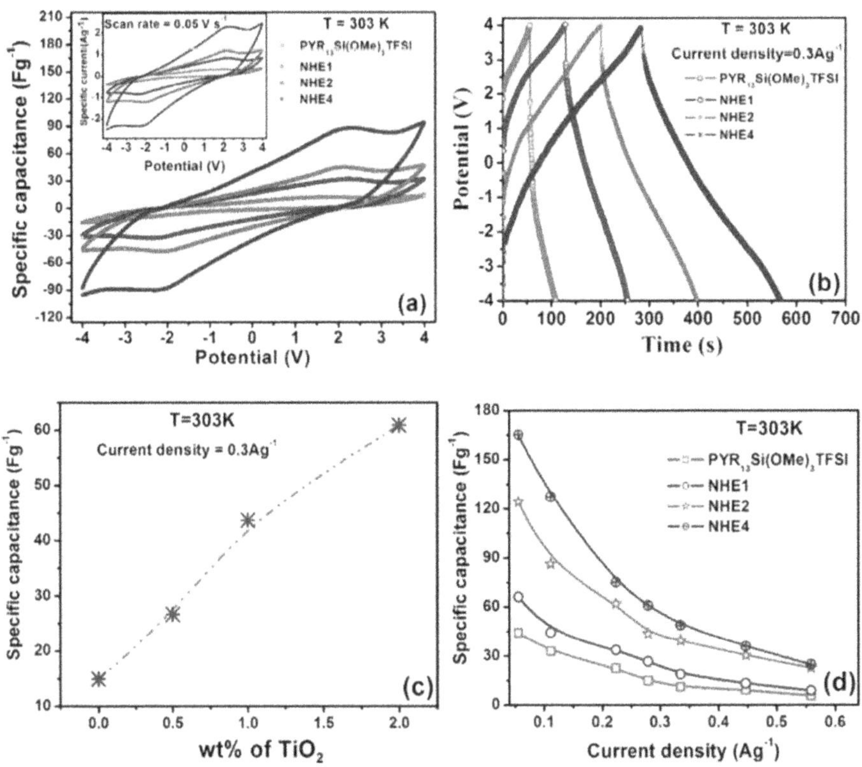

FIGURE 3.13 (a) Capacitance versus voltage plots measured at a scan rate of 0.05 V s^{-1} for a symmetric graphite-based EDLC comprising different nanohybrid electrolytes. The inset shows the corresponding cyclic voltammetry (CV) curves, (b) galvanostatic charge–discharge (GCD) profiles recorded at a specific current of 0.3 A g^{-1} for cells containing pure IL and hybrid electrolytes at 303K, highlighting their distinctive discharge characteristics, (c) variation of electrode capacitance, extracted from GCD measurements, as a function of TiO_2 content (wt.%) within the electrolytes, and (d) dependence of specific capacitance on current density, demonstrating the rate capability and electrochemical robustness of the developed electrolyte systems. (Reproduced with permission from [54]. Copyright (2017) The Electrochemical Society.)

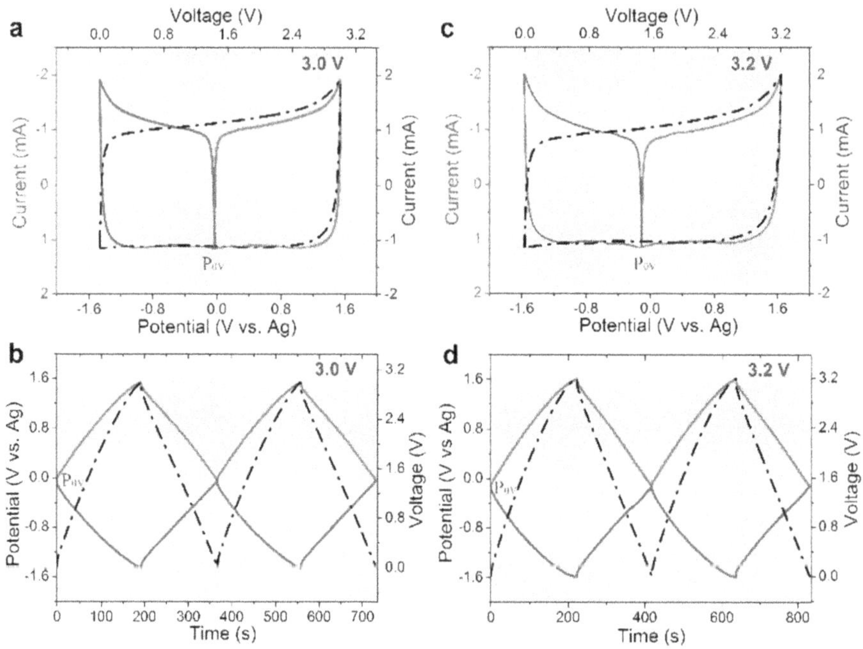

FIGURE 3.14 Electrochemical performance of the EDLC with SiO_2–IL–NTf_2–incorporated electrolyte. CV and GCD curves of the positive AC electrode (blue), negative AC electrode (pink), and the complete EDLC (black dashed) were recorded in the 10% SiO_2–IL–NTf_2-added BMIM–NTf_2/PC electrolyte. Measurements were conducted at operational voltages of 3.0 V (a and b) and 3.2 V (c and d). The potential ranges of the positive and negative electrodes are plotted with respect to the Ag reference electrode. The CV scans were performed at a rate of 10 mV s^{-1}, and the GCD tests were carried out at a current density of 0.5 A g^{-1}, highlighting the voltage-dependent behavior and potential balancing of the two electrodes within the optimized hybrid electrolyte system. (Reproduced with permission from [71]. Copyright (2019) Elsevier B. V.)

at its theoretical operational voltage of 3.2 V, achieving enhanced cycling stability and a 39% improvement in energy density. The same adjustment strategy also improved the cyclic durability of EDLCs utilizing NEt_4–BF_4/PC electrolytes, demonstrating the broader applicability of this approach, as shown in Figure 3.14.

Overall, this strategy presents a promising pathway toward expanding the operational potential window of AC-based EDLCs employing hybrid IL-based electrolytes. The combined improvements in operating voltage, cycling stability, and energy density underscore its potential for advancing the practical viability of high-performance EDLCs.

3.4 CONCLUSIONS

INF electrolytes represent a transformative development in the field of electrochemical storage, overcoming key limitations of conventional IL and organic electrolytes. By combining the tunable physicochemical properties of ILs with the nanoscale functionalities of solid-state additives, these hybrid systems provide enhanced ionic conductivity, superior thermal and electrochemical stability, and mechanical reinforcement. The decoupling of ionic conductivity from viscosity in INF systems enables efficient ion transport even under high energy-density conditions, while the formation of well-structured interfacial EDLs ensures long-term operational stability.

Experimental studies discussed in the chapter demonstrate that INF-based electrolytes effectively mitigate lithium dendrite formation, improve electrode–electrolyte contact, and enhance capacity retention in lithium metal and lithium iron phosphate cells. Similarly, INF-based supercapacitors exhibit expanded voltage windows and improved capacitance retention, validating their potential for high-power and high-temperature operations. Collectively, the synergy between ILs and NPs provides a promising pathway toward safer, durable, and more efficient energy storage devices for next-generation applications such as electric mobility, grid-scale renewable energy storage, and portable electronics.

While INFs offer exceptional potential, real-world integration remains challenging. The high cost and synthesis complexity of ILs, combined with difficulties in maintaining long-term NP dispersion, limit their large-scale use. Research must address issues related to viscosity optimization, recyclability, and environmental impact. Advancements in molecular design and interfacial engineering will be crucial to achieving scalable, eco-friendly hybrid systems. As these materials evolve, INFs may bridge the gap between the safety of solid-state electrolytes and the high performance of conventional liquid systems, defining a new frontier in electrochemical energy storage.

REFERENCES

1. Goodenough JB, Park K-S. The Li-Ion rechargeable battery: A perspective. Journal of the American Chemical Society 2013;135:1167–76.
2. Watanabe M, Thomas ML, Zhang S, Ueno K, Yasuda T, Dokko K. Application of ionic liquids to energy storage and conversion materials and devices. Chemical Reviews 2017;117:7190–239.
3. Kim E, Han J, Ryu S, Choi Y, Yoo J. Ionic liquid electrolytes for electrochemical energy storage devices. Materials 2021, 14(14), 4000; https://doi.org/10.3390/ma14144000.
4. Balducci A. Ionic liquids in lithium-ion batteries. Topics in Current Chemistry 2017;375:20.
5. Zheng Y, Wang D, Kaushik S, Zhang S, Wada T, Hwang J, et al. Ionic liquid electrolytes for next-generation electrochemical energy devices. EnergyChem 2022;4:100075.
6. Bose P, Deb D, Bhattacharya S. Ionic liquid based nanofluid electrolytes with higher lithium salt concentration for high-efficiency, safer, lithium metal batteries. Journal of Power Sources 2018;406:176–84.
7. Armand M, Endres F, MacFarlane DR, Ohno H, Scrosati B. Ionic-liquid materials for the electrochemical challenges of the future. Nature Materials 2009;8:621–9.
8. Yu Z, Tetard L, Zhai L, Thomas J. Supercapacitor electrode materials: Nanostructures from 0 to 3 dimensions. Energy & Environmental Science 2015;8:702–30.
9. Ribeiro A, Lourenço M, de Castro CN. Thermal conductivity of ionanofluids. Proceedings of the 17th Symposium on Thermophysical Properties 2009. p. 21–6.
10. Ishikawa M, Sugimoto T, Kikuta M, Ishiko E, Kono M. Pure ionic liquid electrolytes compatible with a graphitized carbon negative electrode in rechargeable lithium-ion batteries. Journal of Power Sources 2006;162:658–62.
11. Galiński M, Lewandowski A, Stępniak I. Ionic liquids as electrolytes. Electrochimica Acta 2006;51:5567–80.
12. Simon P, Gogotsi Y. Materials for electrochemical capacitors. Nature Materials 2008;7:845–54.

13. Minea AA, Murshed SMS. A review on development of ionic liquid based nanofluids and their heat transfer behavior. Renewable and Sustainable Energy Reviews 2018;91:584–99.
14. Wang W, Wu Z, Li B, Sundén B. A review on molten-salt-based and ionic-liquid-based nanofluids for medium-to-high temperature heat transfer. Journal of Thermal Analysis and Calorimetry 2019;136:1037–51.
15. Joseph A, Sobczak J, Żyła G, Mathew S. Ionic liquid and ionanofluid-based redox flow batteries—A mini review. Energies 2022, 15(13), 4545; https://doi.org/10.3390/en15134545.
16. Deb D, Bhattacharya S. Influence of ionic-liquid-tethered Al_2O_3 nanoparticle on the nonisothermal cold crystallization in ionic-liquid-based nanofluids. The Journal of Physical Chemistry C 2017;121:6962–76.
17. Deb D, Dutta B, Bhattacharya S. Viscosity decoupled charge transport in surface functionalized ZnS nanoparticle dispersed imidazolium ionanofluids. Materials Research Bulletin 2019;116:22–31.
18. Deb D, Bhattacharya S. Ion transport in surface functionalized SnO_2 nanoparticles dispersed imidazolium ionanofluids: Decoupling from structural relaxation. Journal of Molecular Liquids 2019;285:697–706.
19. Deb D, Bhattacharya S. Imidazolium-based ionanofluid electrolytes with viscosity decoupled ion transport properties for lithium-ion batteries. Journal of Molecular Liquids 2023;379:121645.
20. Ray A, Saruhan B. Application of ionic liquids for batteries and supercapacitors. Materials 2021, 14(11), 2942; https://doi.org/10.3390/ma14112942.
21. Mendhe A, Panda HS. A review on electrolytes for supercapacitor device. Discover Materials 2023;3:29.
22. Riedl JC, Akhavan Kazemi MA, Cousin F, Dubois E, Fantini S, Loïs S, et al. Colloidal dispersions of oxide nanoparticles in ionic liquids: Elucidating the key parameters. Nanoscale Advances 2020;2:1560–72.
23. Migowski P, Machado G, Texeira SR, Alves MCM, Morais J, Traverse A, et al. Synthesis and characterization of nickel nanoparticles dispersed in imidazolium ionic liquids. Physical Chemistry Chemical Physics 2007;9:4814–21.
24. Vollmer C, Janiak C. Naked metal nanoparticles from metal carbonyls in ionic liquids: Easy synthesis and stabilization. Coordination Chemistry Reviews 2011;255:2039–57.
25. Łuczak J, Paszkiewicz M, Krukowska A, Malankowska A, Zaleska-Medynska A. Ionic liquids for nano- and microstructures preparation. Part 2: Application in synthesis. Advances in Colloid and Interface Science 2016;227:1–52.
26. He Z, Alexandridis P. Ionic liquid and nanoparticle hybrid systems: Emerging applications. Advances in Colloid and Interface Science 2017;244:54–70.
27. Ninham BW. On progress in forces since the DLVO theory. Advances in Colloid and Interface Science 1999;83:1–17.
28. Hamm SC, Basuray S, Mukherjee S, Sengupta S, Mathai JC, Baker GA, et al. Ionic conductivity enhancement of sputtered gold nanoparticle-in-ionic liquid electrolytes. Journal of Materials Chemistry A 2014;2:792–803.
29. Khatri OP, Adachi K, Murase K, Okazaki K-i, Torimoto T, Tanaka N, et al. Self-assembly of ionic liquid (BMI-PF6)-stabilized gold nanoparticles on a silicon surface: Chemical and structural aspects. Langmuir 2008;24:7785–92.
30. Obliosca JM, Arellano IHJ, Huang MH, Arco SD. Double layer micellar stabilization of gold nanocrystals by greener ionic liquid 1-butyl-3-methylimidazolium lauryl sulfate. Materials Letters 2010;64:1109–12.
31. Pereira J, Souza R, Moita A, Moreira A. Nanofluids and ionic fluids as liquid electrodes: An overview on their properties and potential applications. Processes 2023, 11(11), 3189; https://doi.org/10.3390/pr11113189.
32. Chen F-L, Sun IW, Paul Wang H, Huang CH. Nanosize Copper dispersed ionic liquids as an electrolyte of new dye-sensitized solar cells. Journal of Nanomaterials 2009;2009:472950.

33. Saadmim F, Forhad T, Sikder A, Ghann W, M. Ali M, Sitther V, et al. Enhancing the performance of dye sensitized solar cells using silver nanoparticles modified photo-anode. Molecules 2020, 25(17), 4021; https://doi.org/10.3390/molecules25174021.
34. Chen W-L, Chou H-M, Yang Y-C. Inverse estimation of the unknown base heat flux in irregular fins made of functionally graded materials. International Communications in Heat and Mass Transfer 2017;87:157–63.
35. Kataria J, Devi P, Rani P. Importance of structures and interactions in ionic liquid-nanomaterial composite systems as a novel approach for their utilization in safe lithium metal batteries: A review. Journal of Molecular Liquids 2021;339:116736.
36. Gao J, Ndong RS, Shiflett MB, Wagner NJ. Creating nanoparticle stability in ionic liquid [C4mim][BF4 by inducing solvation layering. ACS Nano 2015;9:3243–53.
37. Cheng P, Yang Y, Huang S. Theoretical insights into the interaction between RunPt13-n (n=4, 7 and 9) clusters and [BMIM]+ based ionic liquids: Effect of anion. Journal of Molecular Graphics and Modelling 2017;74:117–24.
38. Gao J, Mwasame PM, Wagner NJ. Thermal rheology and microstructure of shear thickening suspensions of silica nanoparticles dispersed in the ionic liquid [C4mim][BF4]. Journal of Rheology 2017;61:525–35.
39. Zheng W, Liu X, Yan Z, Zhu L. Ionic liquid-assisted synthesis of large-scale TiO_2 nanoparticles with controllable phase by hydrolysis of $TiCl_4$. ACS Nano 2009;3:115–22.
40. Wang L, Chang L, Zhao B, Yuan Z, Shao G, Zheng W. Systematic investigation on morphologies, forming mechanism, photocatalytic and photoluminescent properties of ZnO nanostructures constructed in ionic liquids. Inorganic Chemistry 2008;47:1443–52.
41. He Z, Alexandridis P. Nanoparticles in ionic liquids: Interactions and organization. Physical Chemistry Chemical Physics 2015;17:18238–61.
42. Wittmar A, Ruiz-Abad D, Ulbricht M. Dispersions of silica nanoparticles in ionic liquids investigated with advanced rheology. Journal of Nanoparticle Research 2012;14:651.
43. Pereira J, Souza R, Moreira A, Moita A. An overview of the ionic liquids and their hybrids operating in electrochemical cells and capacitors. Ionics 2024;30:4343–85.
44. Zaccarelli E, Poon WCK. Colloidal glasses and gels: The interplay of bonding and caging. Proceedings of the National Academy of Sciences 2009;106:15203–8.
45. Ueno K, Sano Y, Inaba A, Kondoh M, Watanabe M. Soft glassy colloidal arrays in an ionic liquid: Colloidal glass transition, ionic transport, and structural color in relation to microstructure. The Journal of Physical Chemistry B 2010;114:13095–103.
46. Stokes JR, Frith WJ. Rheology of gelling and yielding soft matter systems. Soft Matter 2008;4:1133–40.
47. Dou Q, Liu L, Yang B, Lang J, Yan X. Silica-grafted ionic liquids for revealing the respective charging behaviors of cations and anions in supercapacitors. Nature Communications 2017;8:2188.
48. Moganty SS, Jayaprakash N, Nugent JL, Shen J, Archer LA. Ionic-liquid-tethered nanoparticles: Hybrid electrolytes. Angewandte Chemie International Edition 2010;49:9158–61.
49. Moganty SS, Srivastava S, Lu Y, Schaefer JL, Rizvi SA, Archer LA. Ionic liquid-tethered nanoparticle suspensions: A novel class of ionogels. Chemistry of Materials 2012;24:1386–92.
50. Lu Y, Moganty SS, Schaefer JL, Archer LA. Ionic liquid-nanoparticle hybrid electrolytes. Journal of Materials Chemistry 2012;22:4066–72.
51. Bose P, Bhattacharya S. Influence of anion structure on thermal, mechanical, dielectric and electrochemical properties of ZrO_2 nanoparticle-pyrrolidinium ionic liquid hybrid electrolytes for the application in energy storage devices. International Journal of Hydrogen Energy 2018;43:4090–100.
52. Dutta B, Deb D, Bhattacharya S. Ionic liquid-SnO_2 nanoparticle hybrid electrolytes for secondary charge storage devices: Physicochemical and electrochemical studies. International Journal of Hydrogen Energy 2018;43:4081–9.

53. Deb D, Bhattacharya S. Role of different nanoparticulate cores on the thermal, mechanical and electrochemical cycling behaviour of nanoscale hybrid ionic fluids. Electrochimica Acta 2017;245:438–47.
54. Bose P, Bhattacharya S. Electrochemical cycling behavior of pyrrolidinium ionic liquid tethered TiO_2 nanoparticle-hybrid electrolytes: Influence of grafting density. Journal of The Electrochemical Society 2017;164:H788.
55. Fukushima T, Aida T. Ionic liquids for soft functional materials with carbon nanotubes. Chemistry – A European Journal 2007;13:5048–58.
56. Lu Y, Korf K, Kambe Y, Tu Z, Archer LA. Ionic-liquid–nanoparticle hybrid electrolytes: Applications in lithium metal batteries. Angewandte Chemie International Edition 2014;53:488–92.
57. Li Y, Wong K-W, Ng K-M. Ionic liquid decorated mesoporous silica nanoparticles: A new high-performance hybrid electrolyte for lithium batteries. Chemical Communications 2016;52:4369–72.
58. Bose P, Deb D, Bhattacharya S. Lithium-polymer battery with ionic liquid tethered nanoparticles incorporated P(VDF-HFP) nanocomposite gel polymer electrolyte. Electrochimica Acta 2019;319:753–65.
59. Niu H, Wang L, Guan P, Zhang N, Yan C, Ding M, et al. Recent advances in application of ionic liquids in electrolyte of lithium ion batteries. Journal of Energy Storage 2021;40:102659.
60. Karuppasamy K, Theerthagiri J, Vikraman D, Yim C-J, Hussain S, Sharma R, et al. Ionic liquid-based electrolytes for energy storage devices: A brief review on their limits and applications. Polymers 2020, 12(4), 918; https://doi.org/10.3390/polym12040918.
61. Rana S, Thakur RC, Dosanjh HS. Ionic liquids as battery electrolytes for lithium ion batteries: Recent advances and future prospects. Solid State Ionics 2023;400:116340.
62. Wang J, Xu L, Jia G, Du J. Challenges and opportunities of ionic liquid electrolytes for rechargeable batteries. Crystal Growth & Design 2022;22:5770–84.
63. Li Y, Wong KW, Dou Q, Ng KM. A single-ion conducting and shear-thinning polymer electrolyte based on ionic liquid-decorated PMMA nanoparticles for lithium-metal batteries. Journal of Materials Chemistry A 2016;4:18543–50.
64. Chen N, Dai Y, Xing Y, Wang L, Guo C, Chen R, et al. Biomimetic ant-nest ionogel electrolyte boosts the performance of dendrite-free lithium batteries. Energy & Environmental Science 2017;10:1660–7.
65. Li L, Li S, Lu Y. Suppression of dendritic lithium growth in lithium metal-based batteries. Chemical Communications 2018;54:6648–61.
66. Delacroix S, Sauvage F, Reynaud M, Deschamps M, Bruyère S, Becuwe M, et al. SiO_2/ionic liquid hybrid nanoparticles for solid-state lithium ion conduction. Chemistry of Materials 2015;27:7926–33.
67. Lu Y, Das SK, Moganty SS, Archer LA. Ionic liquid-nanoparticle hybrid electrolytes and their application in secondary lithium-metal batteries. Advanced Materials 2012;24:4430–5.
68. Xu S, Lu Y, Wang H, Abruña HD, Archer LA. A rechargeable Na–CO_2/O_2 battery enabled by stable nanoparticle hybrid electrolytes. Journal of Materials Chemistry A 2014;2:17723–9.
69. Zheng S, Wu Z-S, Wang S, Xiao H, Zhou F, Sun C, et al. Graphene-based materials for high-voltage and high-energy asymmetric supercapacitors. Energy Storage Materials 2017;6:70–97.
70. Chi Y-W, Hu C-C, Shen H-H, Huang K-P. New approach for high-voltage electrical double-layer capacitors using vertical graphene nanowalls with and without nitrogen doping. Nano Letters 2016;16:5719–27.
71. Dou Q, Lian C, Lei S, Chen J, Liu H, Yan X. Silica-grafted ionic liquid for maximizing the operational voltage of electrical double-layer capacitors. Energy Storage Materials 2019;18:253–9.

4 Thermophysical Properties of Ionanocolloids

Alina Adriana Minea, Engin Gedik, and Dana Bejan

4.1 INTRODUCTION

Ionanocolloids are at the forefront of developing new heat transfer fluids, as it has been proven that their properties can be tailored for different applications. The concept behind ionanocolloids is similar to that of manufacturing nanocolloids, but with ionic liquids (ILs) as the base fluid. Thus, the manufacturing techniques are similar, as well as the experimental approach in defining their thermophysical properties.

One of the ways to develop energy-efficient systems is to improve heat transfer, which is a fundamental topic of engineering applications and has a wide range from living organisms to industrial systems. In areas where heat transfer is relevant, especially in thermal systems, the working fluid is extremely effective in determining the performance of the system. Working fluids that are habitual and widely used in thermal energy systems are water, oil, and ethylene glycol. These are included in the literature as conventional heat transfer fluids. Many studies have been carried out recently and continue to be carried out to replace conventional heat transfer fluids with fluids with improved thermophysical properties. The focus of these studies is the use of fluids with improved thermal properties instead of conventional heat transfer fluids. Nanofluids are an example of these, which are suspensions formed by adding nano-sized solid particles in certain proportions to base fluids such as water, oil, and ethylene glycol. In general, studies on this subject show that heat transfer improves with the use of nanofluids instead of conventional fluids. This improvement in heat transfer may vary depending on different parameters such as the type of nanofluid used, the production method, the concentration of nanoparticles (NPs) that make up the nanofluid suspension, the type and shape of the NP, and its morphological properties. In addition, suspensions formed with NPs dispersed in ILs have recently emerged as innovative liquids in heat transfer applications, and the development process is summarized in Figure 4.1.

This chapter discusses the thermophysical properties with a focus on thermal conductivity, density, and isobaric specific heat, containing also several insights into the stability and degradation of ionanocolloids.

The discussion will be focused on available techniques, experimental data, as well as proposed correlations, where applicable.

DOI: 10.1201/9781003595250-4

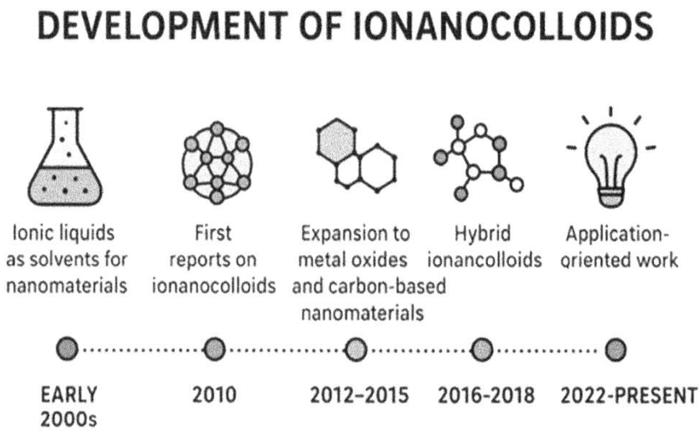

FIGURE 4.1 Development of ionanocolloids.

Figure 4.2 presents the main advantages of ILs, outlining their relevance for possible new heat transfer fluids, while an in-depth analysis is inserted in Minea [1], and in Table 4.1 are several possible applications of ionanocolloids in different areas [1].

In Table 4.2, it is interesting to notice a small review of literature outcomes on several ILs, along with an assessment with one of the most popular heat transfer fluids, such as for example Therminol 68. It needs to be underlined here that data on ILs is rather limited in the open literature, and their employment as a heat transfer is increasing.

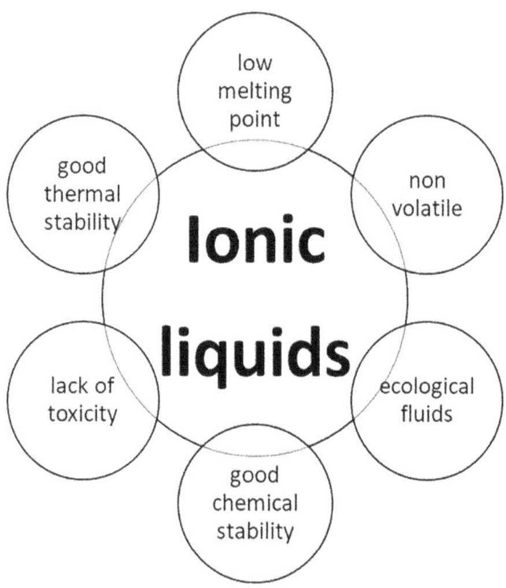

FIGURE 4.2 Advantages of ionic liquids.

TABLE 4.1
Ionanocolloids Possible Applications

Applications of ionanocolloids	Electrolytes	Fuel cells
		Batteries
		Sensors
	Solvents	Bio catalysis
		Organic reactions
		Nanoparticle synthesis
		Polymerization
	Lubricants and additives	
	Separation processes	Gas separation
		Distillation
		Extraction
		Membranes
	Liquid crystal displays	
	Analytics	Solvent
		Protein crystallization
	Electroelastic materials	Robotics
		Artificial muscles
	Heat transfer	

4.2 THERMAL CONDUCTIVITY OF IONANOCOLLOIDS

As can be seen from Table 4.1 where several properties of ILs are inserted, also in terms of thermal conductivity, one can notice that the thermal conductivity of most ILs lies around 0.2 W/m K. Another interesting observation that arises from the published studies (see for example: [1, 15–25])] is that there is little to no influence of temperature over the values of ILs thermal conductivity. This phenomenon arises due to the cation type and structure and is less influenced by the anion type [1].

Thermal conductivity of ionanocolloids was deeply discussed by Minea and Murshed [15, 20] in an attempt to raise attention to these possible new heat transfer fluids, while further on a few insights are added on the effect of temperature and concentration on thermal conductivity variation.

4.2.1 Effect of Temperature on Thermal Conductivity Variation

The temperature variation of ionanocolloids' thermal conductivity is widely studied, as was noticed from the open literature (see [1–45] for further results on thermal conductivity that will be discussed in this section). In Figure 4.3, several experimental data performed are inserted by França et al. [9], who added different concentrations of multiwall carbon nanotubes (CNTs) (MWCNTs) in two ILs: $[C_4mim][(CF_3SO_2)_2N]$ and $[C_2mim][EtSO_4]$. Results revealed that thermal conductivity is influenced less by temperature rise and more substantially by the NPs addition.

TABLE 4.2
Therminol 68 and Ionic Liquids: Properties Comparison

Name		Density, g/cm^3	Isobaric Specific Heat, J/g K	Thermal Conductivity, W/mK	Reference
Therminol 68		0.860 at 293 K	1.880 at 373 K	0.123 at 293 K	[2]
[EMIM][BF_4]	1-Ethyl-3-methylimidazolium tetrafluoroborate	1.218 at 333 K	1.281 at 373 K	0.200 at 293 K	Sanmamed et al. [3]
[BMIM][BF_4]	1-Butyl-3-methylimidazolium tetrafluoroborate	1.181 at 333 K	1.627 at 373 K	0.186 at 293 K	Kakinuma et al. [4]
[BMIM][Tf_2N]	1-Butyl-3-methylimidazolium bis(trifluoromethylsulfonyl)imide	1.384 at 333 K	1.430 at 373 K	0.127 at 293 K	Wadekar [5]
[HMIM][BF_4]	1-Hexyl-3-methylimidazolium tetrafluoroborate	1.145 at 333 K	2.355 at 373 K	0.162 at 293 K	Wang et al. [6]
[C_2mim][CH_3SO_3]	1-Ethyl-3-methylimidazolium methanesulfonate	1.214 at 333 K	1.773 at 333 K	0.188 at 293 K	Cherecheş et al. [7]
[C_4mim][NTf_2]	3-Butyl-3-octylimidazolium bis[trifluoromethylsulfonyl]imide	1.386 at 333 K	1.740 at 333 K	0.126 at 293 K	Paul et al. [8]
[C_2mim][$EtSO_4$]	1-Ethyl-3-methyl imidazoliumethylsulfate	1.241 at 293 K	1.615 at 293 K	0.175 at 293 K	França et al. [9]
[C_4mim][Cl]	1-n-Butyl-3-methylimidazolium chloride	1.086 at 293 K	1.849 at 293 K	0.176 at 293 K	Patil et al. [10] Fredlake [11]
[C_4mim][Br]	1-Butyl-3-methylimidazolium bromide	1.293 at 293 K	1.447 at 293 K	0.160 at 293 K	Patil et al. [10] Fredlake [11]
[C_4mim][I]	1-Butyl-3-methylimidazolium iodide	1.489 at 293 K	1.142 at 293 K	0.131 at 293 K	Patil et al. [10] Yamamuro and Kofu [12]
[C_4mim][BF_4]	1-Butyl-3-methylimidazolium tetrafluoroborate	1.210 at 293 K	1.612 at 293 K	0.163 at 293 K	Ribeiro et al. [13] Paulechka et al. [14]

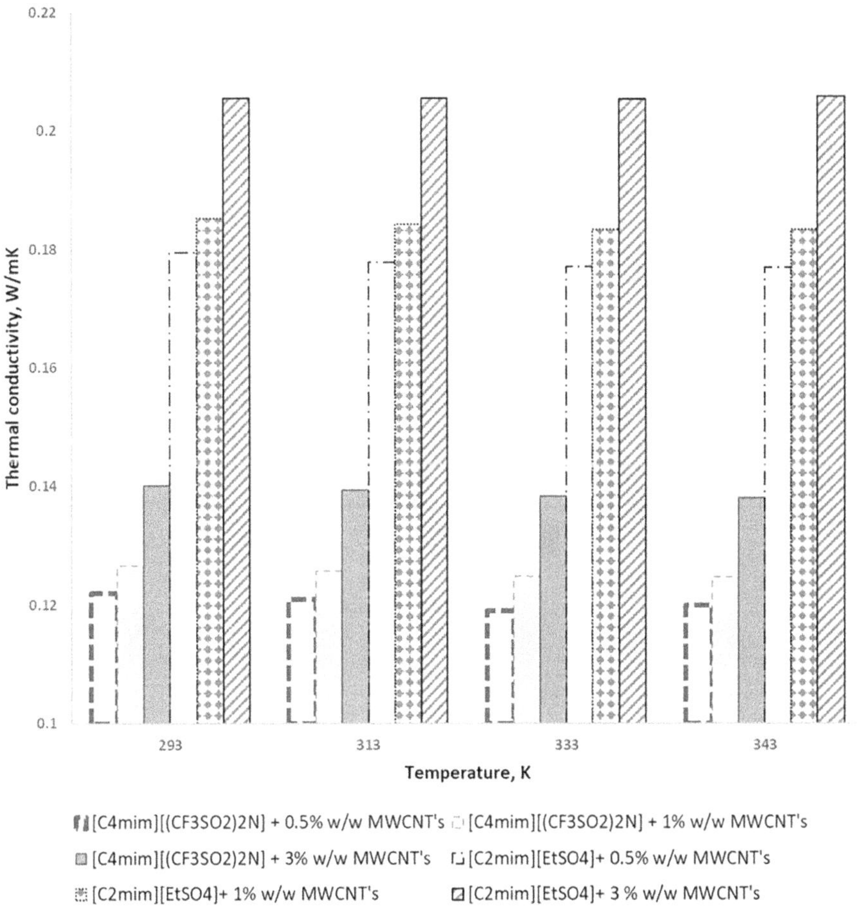

FIGURE 4.3 Thermal conductivity of ionanocolloids at different temperatures.

Ribeiro et al. [13, 27] employed also MWCNTs and used three ILs as a base. The experimental data, as shown in Figure 4.4, also revealed a low influence of the temperature on the augmentation of thermal conductivity. Plus, the authors found that the thermal conductivity of employed ILs diminished linearly in the considered temperature array, a phenomenon acknowledged in the literature for molecular liquids, as well as ILs [1].

Paul [8] investigated the thermal conductivity of several ILs, such as [C_4mpyrr][NTf_2], [C_4mim][NTf_2], [N_{4111}][NTf_2], as well as their nanocolloids with alumina. The experiment outlined the influence of alumina NPs on thermal conductivity, and some of the outcomes are inserted in Figure 4.5. The fallouts confirmed a habitually small increase in thermal conductivity of the investigated nanocolloids when the temperature rises.

Wang et al. [6] manufactured [HMIM] [BF_4]+graphene and MWCNT and noticed a minimum increase in thermal conductivity when temperature rises, as well as when NPs are added to the IL, as can be noticed from Figure 4.6.

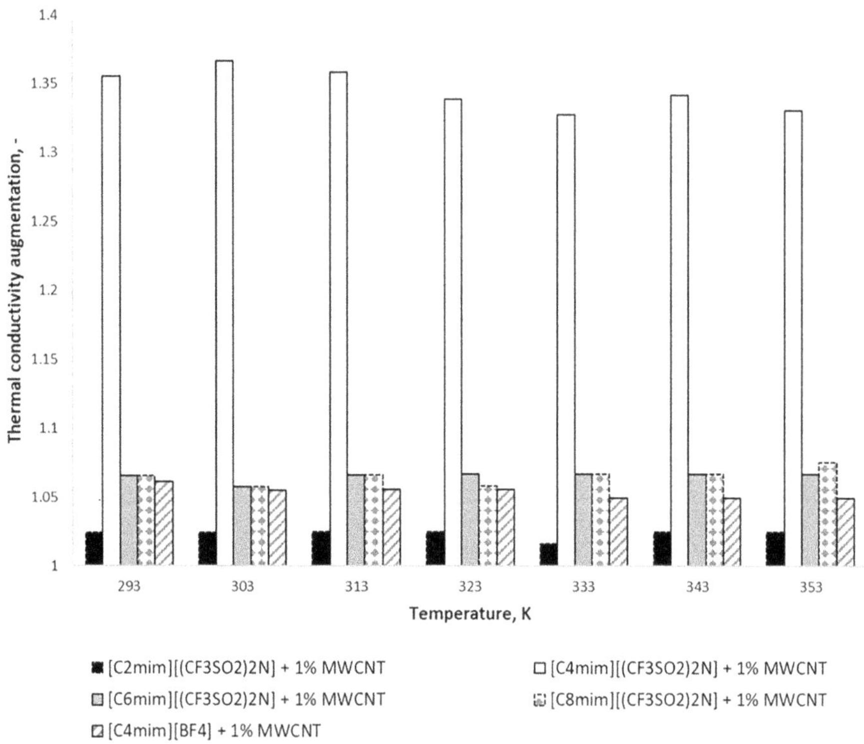

FIGURE 4.4 Thermal conductivity augmentation versus temperature for several ILs + MWCNT.

Sundar et al. [41] employed nanodiamond/copper oxide (ND/CuO) NPs dispersed into 1-ethyl-3-methylimidazolium chloride [EMiM][Cl] to create ionanocolloids and analyzed the thermal conductivity under particle volume loadings from 0.2 to 1.0%, and temperature up to 333 K. Compared to the IL, the thermal conductivity of the ionanocolloids increased by 35.20% at 1.0 vol.% of ionanocolloids and at a temperature of 333 K. On the other hand, the addition of 0.5% graphene oxide (GO) in the same IL [42] enhanced thermal conductivity by 26.39% at the same temperature.

Moosavi et al. [43] discussed thermal conductivity of [HMIM][BF4] IL + SiC NPs in the temperature range of 298.15–338.15 K and at different NP volume fractions, i.e. 1.10, 2.27, and 3.47%. The results are in line with the open literature, and an increase in thermal conductivity was attained both by adding NPs and increasing temperature.

4.2.2 Concentration Influence on Ionanocolloids Thermal Conductivity

The addition of solid NPs to any fluid increases the thermal conductivity, and this phenomenon is widely discussed in the literature. As for basic nanofluids, ionanocolloids reveal the same behavior, while part of the outcomes will be discussed in the next lines.

França et al. [9] added 0.5–3 wt.% MWCNT into two ILs (i.e. [C_4mim][$(CF_3SO_2)_2N$] and [C_2mim][$EtSO_4$]) and reported thermal conductivity at ambient

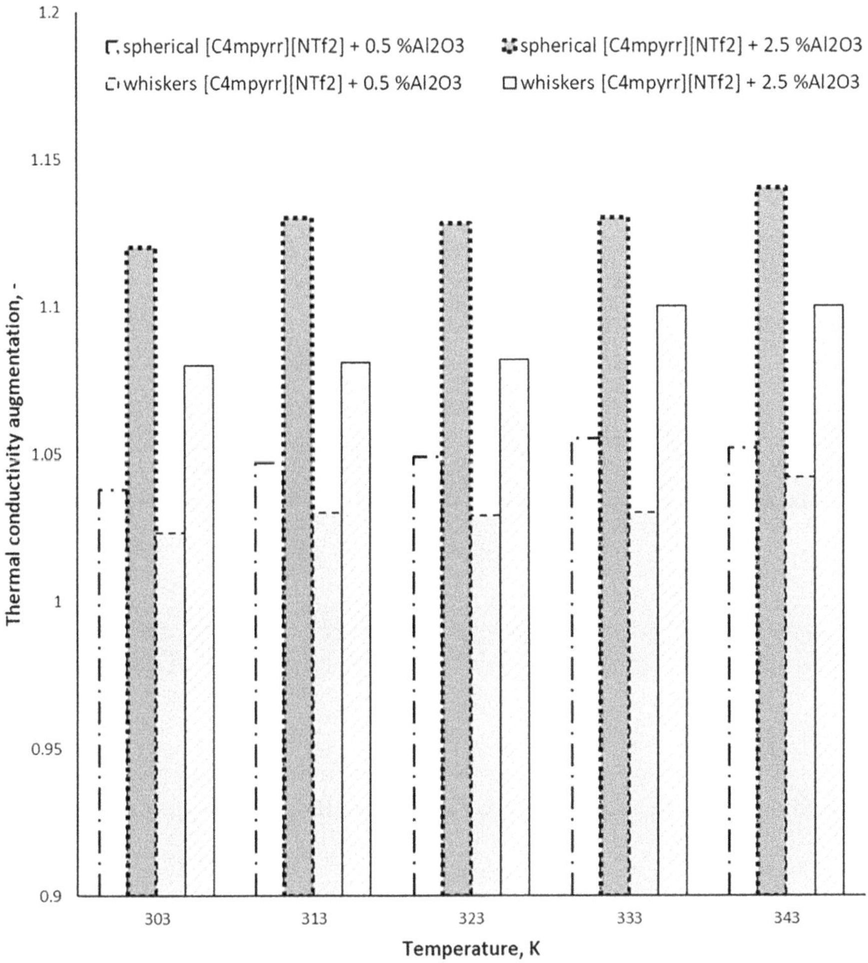

FIGURE 4.5 Thermal conductivity augmentation with temperature for nanocolloids with alumina.

temperature (see Figure 4.7). From the reported experimental data, one can notice that thermal conductivity increased when the NPs loading increased.

Effect of NP loading and shape of Al_2O_3 (<50nm) on the ionanocolloids thermal conductivity was reported by Paul [8] (see Figure 4.8), and the authors noticed that the thermal conductivity noticeably increased when the alumina concentration increased.

Results from Nieto de Castro et al. [30] are exemplified in Figure 4.9, where it can be seen that the increase in thermal conductivity is moderate.

Figure 4.10 presents several data extracted for MWCNT with a number of ILs, as were outlined by França et al. [9], Wang et al. [6], and Xie et al. [46]. Plus, it can be easily seen from Xie et al. [46] that the thermal conductivity enhancement is lower when using a mixture of IL and water.

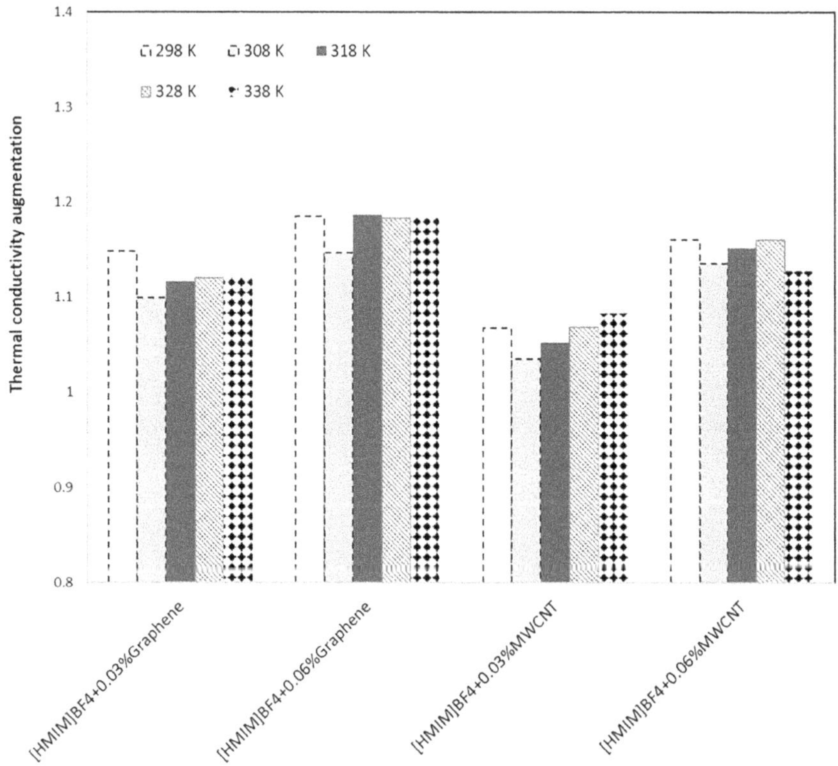

FIGURE 4.6 Augmentation of thermal conductivity for [HMIM][BF_4] + graphene/MWCNT.

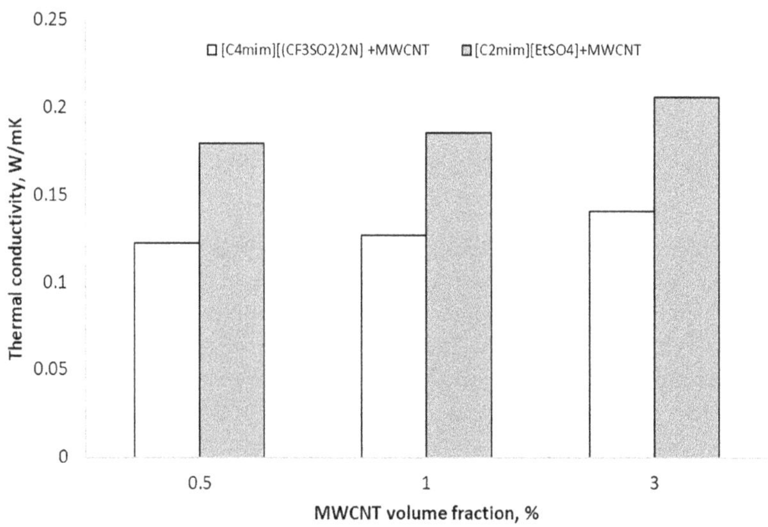

FIGURE 4.7 Thermal conductivity variation of two ionanocolloids versus MWCNT volume fraction.

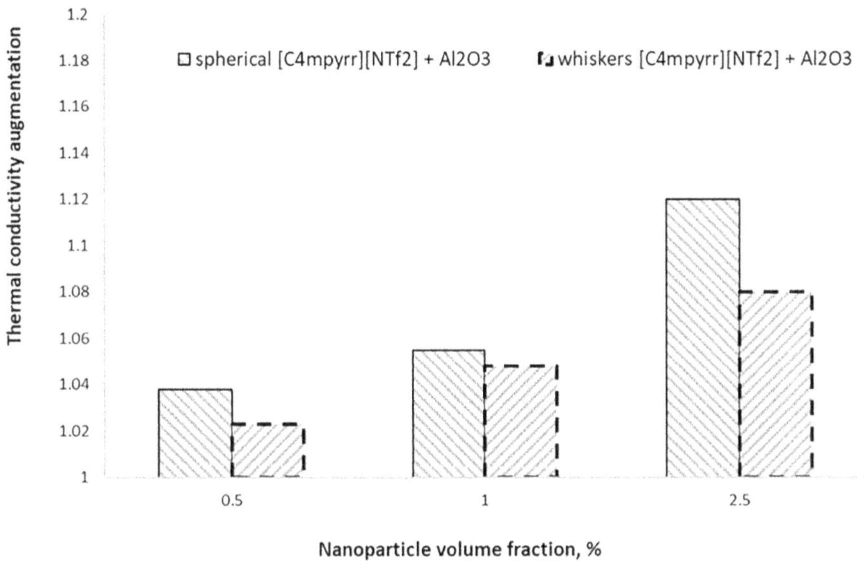

FIGURE 4.8 Thermal conductivity for [C_4mpyrr][NTf_2]+spherical Al_2O_3 and [C_4mpyrr][NTf_2] + whiskers Al_2O_3 at 313 K.

On the other hand, Wang et al. [6] noticed a larger increase with little addition of NPs (i.e. 0.03 or 0.06 wt.% MWCNT). This phenomenon may be due to different equipment employed in measuring the thermal conductivity, or it can be reflected by the purity of the base fluid.

Figure 4.11 also presents several experimental data on [C_4mpyrr][NTf_2], [C_4mim][NTf_2], [C_2mim][CH_3SO_3], and a mixture between [C_2mim][CH_3SO_3] and water

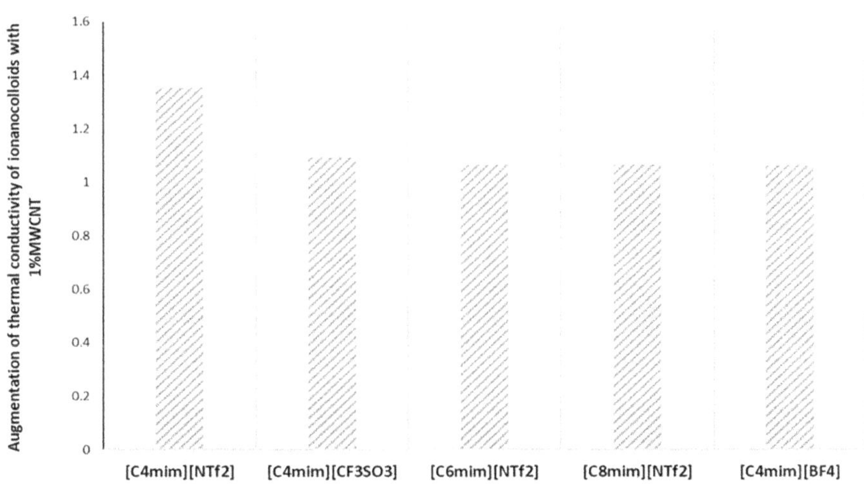

FIGURE 4.9 Thermal conductivity of several MWCNT ionanocolloids.

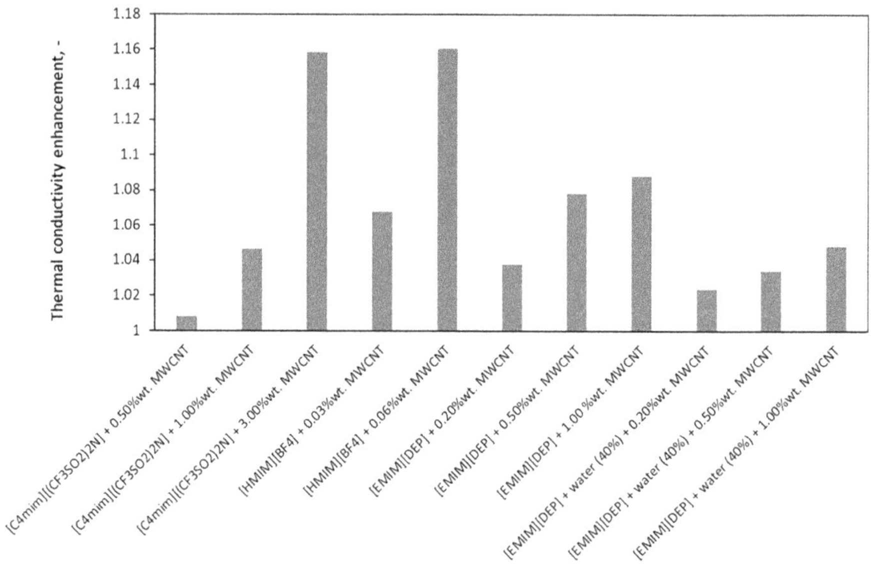

FIGURE 4.10 Base fluid influence on MWCNTs ionanocolloids thermal conductivity values.

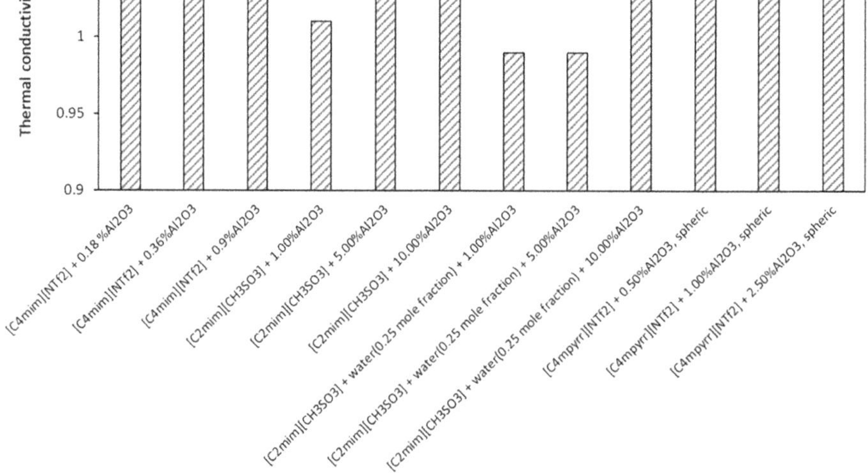

FIGURE 4.11 Base fluid influence on alumina ionanocolloids thermal conductivity.

(with 0.25 mole fraction) enhanced with alumina NPs [21–24]. Outcomes revealed that a loading of 1 wt.% alumina results in to a thermal conductivity augmentation of up to 10%, revealing the importance of the base fluid thermophysical properties.

Additionally, in Figure 4.12 are plotted the data of [HMIM][BF4] ionanocolloids thermal conductivity using NPs as: graphene, MWCNT, SiC, as discussed by Minea and Murshed [15, 20]. The experimental results reveal the influence of NP type on the thermal conductivity, which is a normal phenomenon, due to the different NPs' properties. Nevertheless, the influence of NP size was not scrutinized in the open literature.

Other experiments were performed on water IL mixtures by Chereche̦s et al. [21–23], and the results showed an increase in thermal conductivity of suspensions when alumina was added to the water+[C_2mim][CH_3SO_3] IL, as can be seen in Table 4.3. In Table 4.3, the manufactured mixtures can also be seen, in percentages of mole fraction.

4.2.3 Key Mechanisms of Thermal Conductivity Enhancement

Concluding, most of the research clearly outlined an increase in thermal conductivity with both temperature and NP loading. The key mechanisms can be summarized as follows:

1. ***Brownian motion*** is the random movement of NPs within the base fluid caused by collisions with the fluid molecules. This constant motion has two main effects:
    - *Micro-convection*: The movement of NPs creates microscopic fluid currents (nanoscale convection) that help transport heat more effectively than in a static fluid. This is considered a significant factor, especially at higher temperatures where particle movement is more vigorous.
    - *Particle–fluid interaction*: The constant bombardment and interaction between the moving NPs and the base fluid molecules facilitate more efficient energy transfer.
2. ***NP clustering*** refers to the aggregation of NPs into chain-like or fractal structures. This is a complex phenomenon with both positive and negative impacts on thermal conductivity.
    - *Improved thermal pathways*: When NPs cluster, they form low-resistance thermal pathways that can conduct heat much faster than the surrounding base fluid. This is particularly effective if the clusters form a percolating network, providing a continuous path for heat transfer.
    - *Reduced effectiveness*: Excessive clustering can lead to sedimentation and uneven particle distribution, which can actually decrease the overall thermal conductivity and system stability.
3. ***Nano-layering*** is the formation of a thin, ordered layer of base fluid molecules at the surface of the NPs. The properties of this layer are different from those of the bulk fluid.

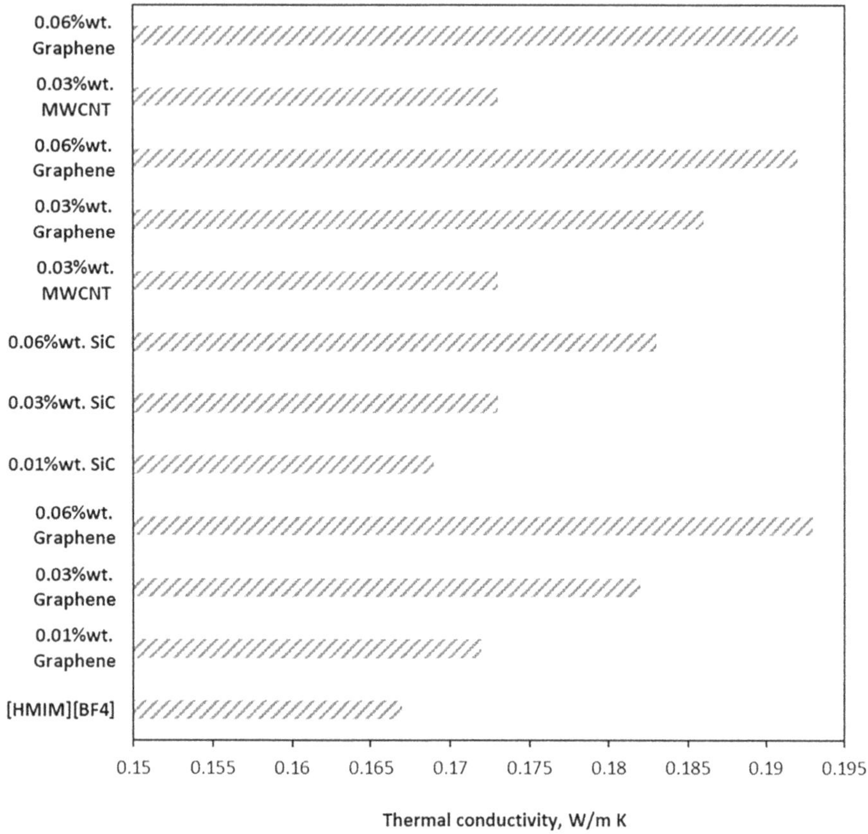

FIGURE 4.12 Ionanocolloids based on [HMIM][BF4] – influence of the nanoparticle type.

TABLE 4.3
Thermal Conductivity of Water, Ionic Liquid and Ionanocolloids at Ambient Temperature (Where IL Refers to [C_2mim][CH_3SO_3]) [21–23]

Fluids	Ionanocolloids with Alumina	Thermal Conductivity (W/m K)
Water		0.598
[C_2mim][CH_3SO_3] IL		0.188
0.25W+0.75IL		0.201
0.50W+0.50IL		0.201
0.75W+ 0.25IL		0.229
	1 wt.% Al_2O_3/(0.25W+0.75IL)	0.195
	2.5 wt.% Al_2O_3/(0.25W+0.75IL)	0.197
	5 wt.% Al_2O_3/(0.25W+0.75IL)	0.201
	10 wt.% Al_2O_3/(0.25W+0.75IL)	0.210
	15 wt.% Al_2O_3/(0.25W+0.75IL)	0.222

- *Thermal bridge*: This solid-like nanolayer acts as a thermal bridge, improving heat transfer between the solid NP and the liquid base fluid. Its thermal conductivity is typically higher than the bulk fluid but lower than the NP itself.
- Thermal Boundary Resistance is the resistance to heat flow at the interface between two materials, in this case, the NP and the base fluid.
- *Interfacial resistance*: While the other mechanisms enhance heat transfer, Kapitza resistance can hinder it. The larger the surface area-to-volume ratio of the NPs, the more significant this resistance becomes. However, the high surface area of NPs also increases the number of contact points for heat transfer, which can counteract this resistance.

In summary, the ionanocolloids' improved thermal conductivity is not due to a single cause but rather the interplay of these complex, often competing, mechanisms. The relative importance of each factor depends on the specific fluid properties, such as particle type, size, shape, and concentration, as well as the temperature.

Finally, the phenomenon of thermal conductivity increase is comparable with that observed for all NP enhanced liquids, while the Brownian motion appears to be the main acknowledged mechanism (see the reviews published by Minea [1] and Minea and Murshed [15, 20]). Future research should focus on clarifying how the type of IL and the NP type and shape influence these fluids' thermal properties. This will help in designing more effective heat transfer fluids.

4.3 SPECIFIC HEAT AND DENSITY OF IONANOCOLLOIDS

Thermophysical properties such as specific heat and density, which have a direct impact on issues such as heat transfer, energy efficiency, and energy storage, are critical for system design. These properties, which determine the heat storage ability of a material and the rate at which it can transmit heat, are very effective in the efficient operation of thermal systems. Specific heat (c) is one of these properties and is expressed as the amount of energy that must be given to increase the temperature of a substance by 1 degree. If the event takes place at a constant volume, it is symbolized as c_v, and if it occurs at constant pressure, it is symbolized as c_p, "specific heat". High specific heat allows for greater heat storage and slower temperature change, while providing stability and efficiency in heat transfer processes [15, 25, 40–45, 47–50].

Another parameter that is important in heat transfer processes is density, which is expressed as the mass per unit volume of a material. The high density allows more energy to be stored per unit volume [51, 52]. This phenomenon can provide significant advantages for energy storage applications, especially with the use of phase-change materials and salt hydrates in thermal systems such as solar energy applications. In particular, the material's resistance to temperature fluctuations and changes is important for long-term energy storage. In summary, the effects of specific heat and density on heat transfer performance are presented in Table 4.4.

There are issues to be considered in determining and/or improving thermophysical properties such as specific heat and density. These may vary depending on the purpose of the application in which the material will be used, energy efficiency, and

TABLE 4.4
Effects of Specific Heat and Density Properties on Heat Transfer

Parameter	Effect on Heat Transfer	References
High ρc_p	High heat storage, slow temperature change	[53–55]
Low ρc_p	Low heat storage, rapid temperature change	[53–55]
High density	Slow response, more stable temperature	[54, 56]
High specific heat	Slow response, more heat storage	[53, 54, 56]

cost factors. For example, if heat storage is prioritized in applications, high specific heat and density are preferred. If rapid heat transfer is the precursor, materials with lower specific heat and density can be selected [57].

4.3.1 Specific Heat

Specific heat (*c* [J/(kg K)]) was given in the preceding paragraphs, where it refers to the amount of energy required to increase the temperature of a unit mass of a substance by 1°C (or 1 K). This is a fundamental physical property that indicates how sensitive a substance is to heat. This property is obtained by dividing the heat capacity of a substance (C [J/K]) by the unit mass (m [kg]) (c = C/m). This value is defined in two different ways: at constant pressure (c_p) and at constant volume (c_v). The main difference between them is whether the volume of the system changes under the conditions in which heat is supplied. In c_p, the pressure is kept constant while heat is supplied to the system, the volume can vary, and it is generally easier to measure experimentally and is used more frequently in practice. In c_v, on the other hand, in the volume is kept constant while heat is supplied to the system, the pressure may change. It is especially important in theoretical calculations for solids and liquids.

4.3.1.1 Factors Affecting Specific Heat on Ionanocolloids

Ionanocolloids are innovative fluids used in heat transfer applications, and environmental variables such as temperature and pressure have significant effects on the specific heat of these fluids. Apart from temperature and pressure, there are many other parameters that affect the specific heat of ionanocolloids. Molecular structure, NP type, NP size, and concentration ratio are some of them. However, most of the research also reveals that the effect of these parameters on thermal capacity may vary according to the system under study. The most obvious effect is seen in NP concentration, while the effect of type and size is usually more limited or variable from system to system. For example, with the addition of 0.5–9% MWCNT to the IL [EMIM][BF4], an increase of 0.417–7.99% in thermal capacity can be seen [58]. It has been reported by Oster et al. [59] that the specific heat of ionanocolloids formed by dispersing NPs such as MWCNT, TiO_2, and Al_2O_3 into the same IL increases more than conventional nanofluids. It can be stated that the mechanisms underlying these increases in specific heat are based on electrostatic and dispersive interactions

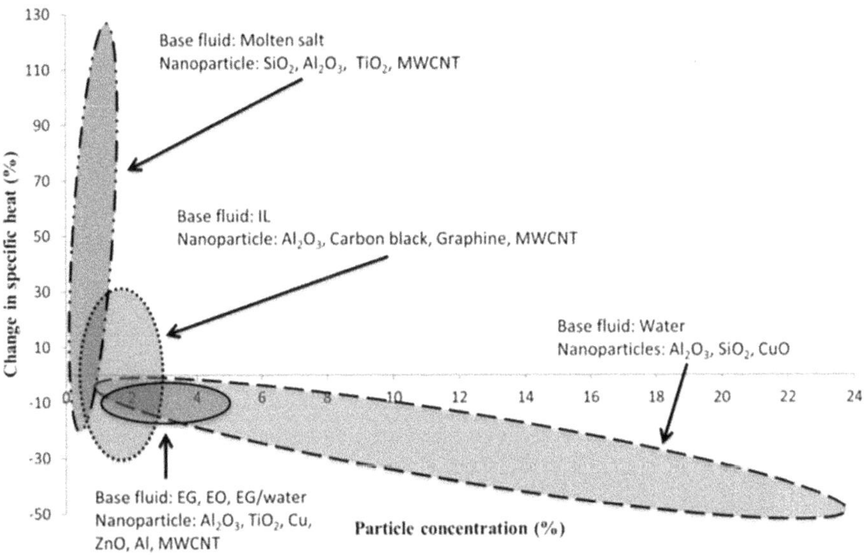

FIGURE 4.13 Summary of specific heat change of different base fluids with adding nanoparticles [63, 64]. (Reprinted with permission from Elsevier.)

between IL and NPs, structured layers formed at the liquid–NP interface, and heat transport dynamics.

As the ***NP concentration ratio*** increases, most of the ionanocolloids exhibit behavior similar to that of nanofluids, resulting in a decrease in the specific heat value. This is because the specific heat value of the NPs is lower than that of the base liquid that forms the suspension [60]. However, although it has been stated in some studies [61, 62] that there is an increase in the specific heat value, especially at very low NP concentrations, it has been concluded that this effect disappears with the increase in the concentration rate, and the situation reverses. When Figure 4.13 is examined, it can be seen that both increases and decreases occur in specific heat changes due to the increase in NPs of ionanocolloids formed by ILs. The specific heat changes of not only ionanocolloids but also other types of nanofluids can be clearly seen from the figure.

NP type and morphology also have important effects on specific heat. Different types of NPs (e.g. Al_2O_3, SiO_2, CNTs, MgO) have different effects on specific heat. CNTs with a high surface area can significantly increase thermal conductivity and produce very small changes in the specific heat value (usually less than 3%). It has been stated in the literature that some NPs, such as MgO, cause a decrease in the specific heat value in the range of 5–22% [65–67].

The chemical structure and temperature of the IL is are other important factors affecting the specific heat of ionanocolloids. The anion and cation structures of the IL and the alkyl chain length play an important role in determining the specific heat. ILs with a long alkyl chain structure can have a higher specific heat value. In addition, the specific heat of ILs tends to increase as the temperature increases, and many studies confirm this phenomenon [60, 68–70]. Pressure, on the other hand, generally affects the specific heat through the phase and density of the fluid. Because they have low vapor pressure, ionanocolloids are able to maintain their stability under high temperature and pressure conditions [71].

There are also effects of ***agglomeration and NP surface functional groups*** on the specific heat of ionanocolloids. Studies have shown that these effects can produce small changes in specific heat in some cases rather than large changes. For example, functional groups such as carboxyl (–COOH), hydroxyl (–OH), long alkyl chain, or PEG added to the surface of CNTs were studied on the specific heat of ionanocolloids, and in most cases, the addition of these functional groups did not lead to a significant change in the specific heat; the observed changes were generally less than 3%. Although these functional groups have a limited effect on specific heat, they can increase the stability and dispersion of NPs. Theoretically, agglomeration can affect specific heat due to the fact that it reduces the surface area and interaction of particles. However, it has been reported in experimental studies on the subject that agglomeration does not have a significant effect on specific heat, and it has been understood that the changes in specific heat are generally negligible [72].

4.3.1.2 Specific Heat Measurement

Specific heat measurements are very important to determine the performance of ionanocolloids in energy storage and heat transfer applications. Generally, these measurements are carried out depending on temperature and NP concentrations, and experimental correlations are created using the data obtained as a result of the measurement. The specific heat of ionanocolloids can be measured at certain temperature ranges and different concentrations of NPs, usually with devices such as differential scanning calorimetry (DSC). During the measurements, the specific heat values of ionanocolloids formed by adding IL and NPs are compared. Specific heat values may increase depending on the increase in temperature and NP concentration. For this reason, measurements are often repeated at different temperatures and concentrations. The type of IL and NP used will directly affect the specific heat measurement results [6, 25, 60]. For example, the isobaric specific heat of some ILs formed using aluminum NPs according to different temperatures was determined experimentally by Chereches et al. [18] with DSC and is given in Figure 4.14. As can be seen from the figure, the isobaric specific heat increased linearly with temperature. Similarly, it was observed in the experimental study conducted by Kanti et al. [69] that the specific heat of ionanocolloids increased with the increase in temperature. It is also stated in the study that the increase in the concentration rate causes a decrease in the specific heat.

Figure 4.15 summarizes this situation. As a result of some studies carried out in the literature, the correlations created to determine the specific heat of ionanocolloids are shown in Table 4.5, and Table 4.6 presents some studies on the specific heat changes of ionanocolloids containing different NPs.

4.3.1.3 Key Mechanisms for Specific Heat Variation

The specific heat capacity of nanofluids generally varies with the addition of NPs. In most cases, the specific heat of a nanofluid is lower than that of the base fluid, which

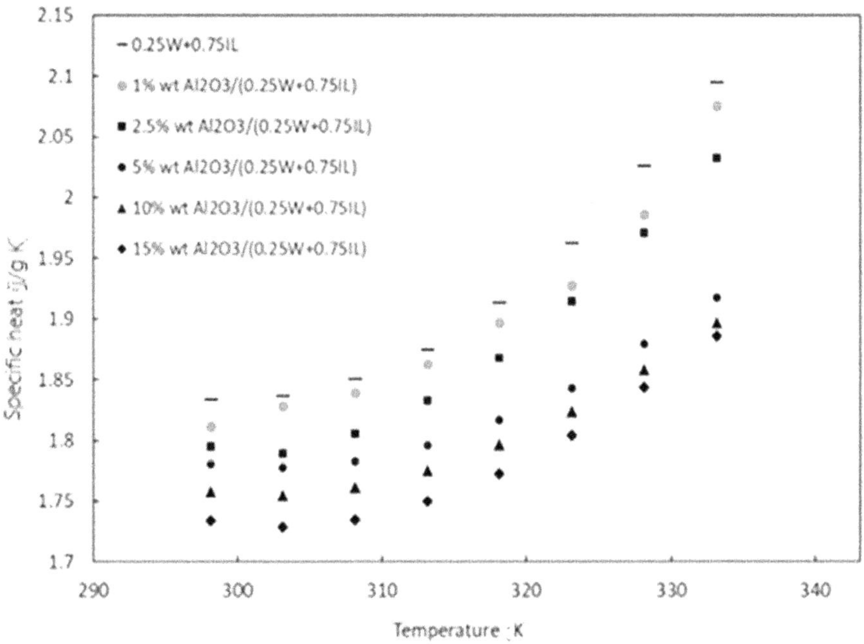

FIGURE 4.14 Variation of specific heat with temperature [18]. (Reprinted with permission from Elsevier.)

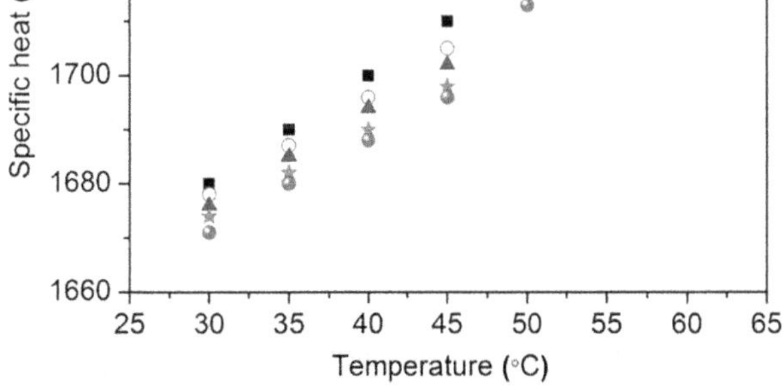

FIGURE 4.15 Specific heat of ionanocolloids with temperature [69]. (Reprinted with permission from Elsevier.)

TABLE 4.5
Correlations for Ionanocolloids Specific Heat

Correlations	Nanoparticle Type	Ionic Liquid	References
$c_{p_{inc}} = \varphi_{wt}\left(c_p\right)_p + \left(1-\varphi_{wt}\right)\left(c_p\right)_{il}$	Al_2O_3 1.0–15.0 wt.%	Mix. water and $[C_2mim][CH_3SO_3]$	[18]
$c_{p_{inc}} = \dfrac{\varphi_{np}\rho_{np}c_{p_{np}} + \varphi_{il}\rho_{il}c_{p_{il}}}{\varphi_{np}\rho_{np} + \varphi_{il}\rho_{il}}$	GO 0.5, 1.0%, ve 2.0 wt.% Al_2O_3 0.5, 1.0%, ve 2.0 wt.%	$[Bmim][NTf_2]$	[73, 74]
$c_{p_{inc}} = \dfrac{\varphi\rho_{np}c_{p_{np}} + \left(1-\varphi\right)\rho_{il}c_{p_{il}}}{\varphi\rho_{np} + \left(1-\varphi\right)\rho_{il}}$	Al_2O_3 0.05–10 wt.% GE 0.05–0.5 wt.%	$[C_2mim][CH_3SO_3]$ [EMIM][Cl]	[16, 69]

ve: volume equivalent

can be a drawback for heat storage applications [40–42]. The mechanisms behind this variation are not fully understood, but a few key factors are believed to play a role.

The most straightforward and widely accepted explanation for specific heat variation is based on the rule of mixtures. This theoretical model suggests that the specific heat of the nanofluid is a weighted average of the specific heat capacities of the NPs and the base fluid. Since most of the NPs used in ionanocolloids typically have a much lower specific heat than the base fluid, adding them to the IL lowers the overall specific heat of the sample.

On the other hand, the degree to which the specific heat changes is directly related to the concentration of the NPs. The more NPs are added, the greater the reduction in specific heat. The type of NP also plays a significant role, as each material has its own intrinsic specific heat [44].

More complex, and still debated, mechanisms involve the interaction between the NPs and the base fluid, for example:

- *Nano-layering*: A layer of ordered fluid molecules may form around the surface of the NPs. This layer can have different properties, including a different specific heat, than the bulk fluid.
- *NP clustering*: Aggregates or clusters of NPs can form within the fluid. The formation and structure of these clusters can affect the overall thermal properties, though the exact impact on specific heat is not as well established as their effect on thermal conductivity.

While these mechanisms are being investigated, the primary reason for the observed variation (i.e. decrease or increase in some cases) in specific heat is the fundamental difference in specific heat between the base fluid and the NP material. This makes a tradeoff between enhanced thermal conductivity and reduced specific heat in some suspensions, which is an important consideration in designing new heat transfer fluids.

TABLE 4.6
Some Examples of Studies in the Literature on Specific Heat Changes of Ionanocolloids

Ionic Liquid	Nanoparticle	Concentration	Explanation	References
[MMIM][DMP]+DW (20:80 vol.%)	MXene (Ti_3C_2)	0.05, 010, and 0.20 vol.%	12.0–14.17% enhancement of specific heat was achieved	[64]
[HMIM][BF_4]	MWCNT and graphene (GE)	0.03, 0.06, and 0.09 wt.%	Specific heat measurements of ionanocolloids were carried out in the range of 20–80°C, and it was observed that the specific heat increased as the temperature increased. However, with the addition of nanoparticles, small decreases in specific heat were observed. When GE/[HMIM][BF4] at a concentration of 0.03% was compared with the base fluid, it was determined that there was a 0.93% decrease in specific heat.	[6]
[C_2mim][CH_3SO_3]	Al_2O_3	0.05–10 wt.%	Specific heat measurements were carried out in the temperature range of 283.15–333.15 K. The results show that as the concentration increases, the specific heat also increases, and significant increases of up to 2.5 wt.%. It was determined that the specific heat of ionanocolloids of 1.0, 2.5, and 5.0% wt. increased by 12.0, 13.7, and 13.8%, respectively, according to the base fluid.	[16]
[C_4mim][NTf_2] [C_4mim][CF_3SO_3] [C_6mim][NTf_2] [C_8mim][NTf_2] [C_4mim][BF_4]	Multiwalled carbon nanotubes (MWCNT)	0.05, 1.0, and 3.0 wt.%	The specific heat of temperature- and concentration-dependent ionanocolloids was investigated, and the results showed that the specific heat of MWCNT-ionanocolloids is higher compared to the base ionic liquid and increases significantly with temperature in the range of 333–363 K. Interestingly, the results showed an abrupt increase of up to 8% in the specific heat over a given temperature range, independent of MWCNT loading. On the other hand, the results showed that there was little increase in the specific heat of ionanocolloids with MWCNT concentration.	[26]
[HMIM][BF_4]	GE	0.03 and 0.06 wt.%	Specific heat measurements of GE/[HMIM][BF_4] ionanocolloid were performed up to 488 K. It was determined that the specific heat value of ionanocolloids at concentrations of 0.03 and 0.06% increased by 4.65% and 10.7 wt.%, respectively, as the temperature increased up to 488 K. The specific heat values of 0.03% and 0.06 wt.% ionanocolloids compared to the base fluid decreased by 1.5 and 3.0%, respectively.	[75]

4.3.2 Density

Density is one of the fundamental physical properties of matter and has an important place in science due to both its definition and its role in thermodynamics. Density is defined as the mass of a substance per unit volume and is mathematically expressed as the ratio of mass to volume ($\rho = m / V$). In its physical definition, density is a value specific to each substance and is one of the distinguishing features of matter. It is used in the identification of the substance along with other physical properties such as color, odor, and melting point. Thermodynamically, it is a fundamental parameter in understanding the behavior of matter, indicating how tightly atoms or molecules are together, and this affects the nature of chemical reactions, biological processes, and atmospheric phenomena. Density significantly affects fluid behavior, energy efficiency, and performance of thermal systems in energy, thermal systems, and heating-cooling processes. When the effects on heat transport are examined, it significantly affects the heat transfer coefficient and the Nusselt number, especially near the critical point or at high temperature differences.

4.3.2.1 Factors Affecting Density in Ionanocolloids

The density of ionanocolloids varies depending on the type, size, and concentration of NPs they contain. The morphology and quantity of NPs significantly affect the physical properties of ionanocolloids, including density. When different carbon-based NPs (e.g. single-walled CNTs – SWCNT, MWCNT) are used, the shape and dimensions of these particles (especially the length/diameter ratio) affect the density of ionanocolloids. Nanotubes with a high aspect ratio are distributed differently in the liquid, leading to changes in macroscopic properties such as density. On the other hand, as the concentration of NPs increases, the density of ionanocolloids also increases. This is because each NP added increases the total mass while the change in volume is less, resulting in an increase in density [66].

The material type of NPs used in ionanocolloids is another important factor that directly affects the density of the system. Since different types of NPs (metal, metal oxide, carbon, etc.) have their own specific density, their ratio and type in the IL also change the total density. Carbon-based NPs generally have a lower density than metals. Therefore, when using low-density NPs such as CNTs, the total density of the ionanocolloids increases less than that of metal or metal oxide NPs. In addition, as the ratio of NPs by weight in the IL increases, the total density of the system also increases. However, this increase varies depending on the density of the NP added.

As important as the effect of NPs in ionanocolloids on density is, the cation and anion structure of the IL, which is the base fluid, especially the molecular weight, directly affects the density. Higher molecular weight usually means higher density. In addition, structural factors such as chain length and branching also change the density [76, 77]. Ionanocolloids are sensitive to physical factors such as temperature and pressure. Research shows that temperature increase, in particular, has a significant effect on density. As the temperature increases, the density of ionanocolloids decreases. Altun et al. [78] reported that in their measurements between 293.15 and 343.15 K, a decrease in density was observed as the temperature increased. The pressure effect, on the other hand, has generally been studied under constant atmospheric conditions, but in general, it is known that the density of liquids increases relatively as the pressure increases.

4.3.2.2 Density Measurement

Density measurement in ionanocolloids is critical to understanding the performance of these materials in heat transfer and energy applications. Density directly affects both the physical properties of the material and the application efficiency. Density is usually measured experimentally. A mixture of IL and NPs is prepared, and density values are recorded at certain temperatures. Density values usually decrease linearly as the temperature increases. However, as the concentration of NPs increases, the density increases. Higher density means more energy-carrying capacity. These changes show how the properties of the material change with the addition of NPs and enable the selection of the most suitable mixture for the application. A density meter can be used to determine the density of ionanocolloids in a wide temperature range. Water is the best option to calibrate the measurement device. With the aid of Pak and Cho's mixture relation, given in Eq. (4.1). density, values can be predicted. Experimental density measurement results (see Figure 4.16) for graphene NP dispersed in 1-ethyl-3-methylimidazolium chloride [EMiM][Cl] IL were presented by Kanti et al. [69], and it was reported in their study that the results obtained were in good agreement with the existing density correlations.

$$\rho_{inc} = \left(1-\varphi_{np}\right)\rho_{il} + \varphi_{np}\rho_{np} \tag{4.1}$$

As can be clearly seen from Figure 4.16(a), with the increase in temperature, the density of the IL and ionanocolloids of different concentrations decreased. Higher density values were obtained at a higher concentration ratio. The fact that graphene NPs have a greater density has also increased the concentration of ionanocolloids. Likewise, when Figure 4.16(b) graph is examined, the density measurement results agree with the mixture theory (Eq. 4.1) presented by Pak and Cho [79] for density. Some other related studies on density changes of ionanocolloids containing different NPs are presented in Table 4.7.

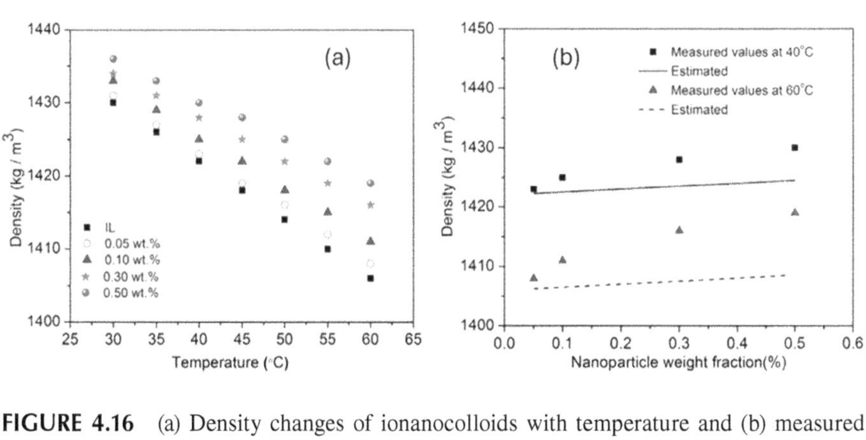

FIGURE 4.16 (a) Density changes of ionanocolloids with temperature and (b) measured density values compared to existing correlation [69]. (Adapted and reprinted with permission from Elsevier.)

TABLE 4.7
Some Examples of Studies in the Literature on Density Changes of Ionanocolloids

Ionic Liquid	Nanoparticle	Concentration	Explanations	References
[EMIM][Cl]	GE	0.05–0.5 wt.%	In the study, density measurements of GE/[EMIM][Cl] ionanocolloid at a concentration of 0.05–0.5 wt.% in the temperature range of 303–333 K were performed. The results showed that density increases with concentration and decreases as temperature increases. The measured density values were compared with the correlations in the literature, and it was determined that the maximum deviation value was 0.74%.	[69]
$[C_4mim][NTf_2]$	Al_2O_3	1 wt.%	In the study, they measured the density of Al_2O_3/[C4mim][NTf2] ionanocolloids with particle sizes of 10, 30, 60, and 90 nm and a concentration of 1.0 wt.% in the temperature range of 293–343 K. The results obtained showed that the density of ionanocolloids does not significantly change with nanoparticle size but tends to decrease as the temperature increases.	[80]
$[P_{14,6,6,6}]$[AcO] $[P_{14,6,6,6}]$[ButO] $[P_{14,6,6,6}]$[HexO] $[P_{14,6,6,6}]$[OctO] $[P_{14,6,6,6}]$[DecO]	MWCNT Graphite (G) Boron nitride (BN) Mesoporous carbon (MC)	0.5, 1.0, 3.0 wt.%	Density measurements of 60 different samples obtained from different ionic liquids and nanoparticles were carried out at temperatures of 298.15 and 363.15 K. The results obtained showed that density increases as the concentration increases. When the values obtained at 298.15 K temperature were compared, it was shown that the highest density value was 3150 kg/m^3 [P14,6,6,6][DecO]+3.0 wt.% RH ionanocolloid.	[81]
$[EMIM][OSO_4]$+Di ethylene glycol	MXene	0.1–0.4 wt.%	It was determined that the density of the samples obtained in the study decreased with temperature increase, which was in good agreement with other studies. On the other hand, it has been determined that the density of the prepared ionic liquid–based nanofluids increases with the addition of nanoparticles due to the high density of MXene.	[82]

4.4 STABILITY AND DEGRADATION OF IONANOCOLLOIDS

4.4.1 Ionic Liquids Stability

In order to use the ILs as base fluids in ionanocolloids, a comprehensive understanding of their thermophysical properties is essential. Consequently, there are multiple criteria for selecting ILs, which are directly determined by their properties. The thermal stability range, non or low volatility, chemical or oxidative stability, viscosity, and polarity are the most important properties that make ILs suitable for selection as media for ionanocolloids. Also, it is important to achieve a balance between what must be addressed in the design of the equipment and the selection of ILs based on associated expenses. Lastly, the environmental impact is a significant criterion in choosing biodegradable ILs whenever possible.

Regarding the interpretation and processing of data for thermodynamic analysis, from the outset, the work conditions must be clearly stated by the authors. An accurate comparison of experimental data with values reported in the literature is required. Thermal analysis of ILs is strongly influenced by a number of critical factors and conditions of measurements: (a) the presence of impurities such as trace water or halide ions; (b) the sample quantity used in measurement; (c) weighing and measuring under dry conditions; and (d) the heating and cooling rate of measurement. These variables can significantly affect the thermal behavior of ILs, as well as their suitability for certain applications. For instance, the presence of water may alter properties like the rate of catalytic reactions or contribute to corrosive effects, and the presence of chloride may alter the viscosity and density value. As an example, in the literature for $[C_2C_1im][TFSI]$, where four different melting points have been reported: 262 K [83], 256 K [11], 258 K [84], and 253 K [34], maybe due to differences in sample purity and preparation conditions.

Therefore, a proper methodology that involves a series of steps must be followed for correct measurements.

a) The detection of halide content, the titration with $AgNO_3$ by ion chromatography [85] or by chloride–sensitive electrodes [86] is required, and for the determination of traces of water, the titration by Karl Fisher is employed. Indicated for ILs is the drying process at high temperature, exceeding 120°C, often accompanied by high vacuum conditions. In the case of some ILs derived from imidazole, it is known that their drying is a challenge, so azeotropic mixtures (dioxane, ethanol) and extended drying periods exceeding 24 hours are used [87]. Detection of these impurities is mandatory for ILs, with recommended levels below 5 ppm; however, researchers indicate that for ILs based on imidazole, the accepted value is below 200 ppm of water or chlorine [88].

b) For thermal transition study (DSC), the sample amount recommended is approximately 4–8 mg [89], and for thermogravimetric analysis (TGA), it is recommended up to 20 mg. The sample amount directly influences the position and the peak intensity, and it was proven that minor shifts in melting point or T_{onset} are expected when the sample quantity varies (not more than 2°C).

c) Regarding the weighing and measuring under dry conditions, it is established that ILs exhibit hygroscopicity, and atmospheric water absorption may significantly influence their properties and applications. For example, the hydrophobicity of ILs containing $[BF_4]^-$anion depends on the alkyl chain length of the associated cation, whereby increasing alkyl chain length from ethyl to octyl increases hydrophobicity. Consequently, weighing and measurements are typically conducted under dry air conditions. It was proved by simulations of mass losses that the curves can be used as an indirect proof of decomposition and implicitly of the evaporation process. Zaitsau et al. [90] proved that, in an open crucible, the temperature dependence of mass loss shows a characteristic behavior of evaporation process, whereas in a closed crucible, the temperature dependence of the mass loss displays a trend that correspond to fluctuations in the reaction rate relative to volume, mass, or quantity of reactant during the transformation with time versus temperature variation.

d) Thermal analysis is a well-established technique used for studying the thermal behavior of substances. Thermogravimetric (TGA) and differential thermogravimetric (DTG) curves are the most used techniques for the characterization of thermal stability and thermal decomposition of pure ILs. Regarding the interpretation of the TGA/DTG analyses, since it is often difficult to compare experimental data over time in the literature, standardized parameters have been adopted for consistent data reporting in this type of analysis. The commonly used parameters for evaluating the thermal decomposition of ILs consist in T_{start} represent the temperature for the start of mass loss, T_{onset} represents the onset decomposition temperature (determined by finding the intersection of the baseline and the tangent to the TGA curve), T_{end} represent the end of decomposition temperature and $T_{0.01/10h}$ represent the temperature at which the IL decomposes 1% over a 10-hour period. An illustrative example is provided in Table 4.8, where multiple values are reported for thermal decomposition of ILs, clearly reflecting the influence of compound purity and the critical factors previously mentioned, which must be explicitly specified in the description of the experimental methodology.

Also, as mentioned before, it is very important to indicate the ratio of the heating, as shown in Table 4.9. Thus, a higher heating rate decreases the heat distribution in molecules, and thermal degradation begins at a higher temperature.

TGA experiments of ILs mostly exhibit one or two distinct thermal decomposition steps. The first weight loss occurred in the range 333–393 K, with a weight loss attributed to the water evaporation and the second weight loss (the most significant mass loss) is associated to the thermal degradation of the anion/cation structures, including several chemical processes including oxidation [101], hydrolysis, dehydration, polymerization, decarboxylation [103, 104]. Thermal decomposition is a chemical property and not a physical property and should be explained in terms of the stabilities of the ionic components of ILs. According to Cao Y. and Mu T., the thermal stabilities of ILs could be classified into five levels [105]. Figure 4.17 provides a summary of the most widely used ILs in the formulation of ionanocolloids, systematically classified into distinct categories: least stable ($T_{onset} < 523$ K); less stable (523 K $\leq T_{onset} < 573$ K); moderately stable (573 K $\leq T_{onset} < 623$ K); more stable with 623 K $\leq T_{onset} < 673$ K interval, and most stable (723 K $\leq T_{onset}$). Anyway, it is noticeable that no observed ILs had thermal stability lower than 373 K, and this aspect is an advantage in choosing ILs for high-temperature applications.

Crucible used in thermal analysis should be an inert material for investigated ILs in order to avoid chemical interaction with temperature variation [106]. Recently, the effect of crucible type (closed, pierced, open) was also discussed by Zaitsau et al. [90],

TABLE 4.8
Comparison of the Thermal Degradation of Some Ionic Liquids Used as Media for Ionanocolloids

Reference	Abbreviation	Rate, K /min	T_{onset}, K	Applications
Parajó et al. [91]	$[C_2C_1im][NTf_2]$	20	724	Thermal fluids
	$[C_3C_1im][NTf_2]$	20	714	Thermal fluids
	$[C_4C_1im][NTf_2]$	20	717	Thermal fluids
	$[C_4C_1C_1im][NTf_2]$	20	715	Thermal fluids
Gómez et al., 2014 [89]	$[C_2C_1im][NTf_2]$	10	692	High specific heat
Niu et al. [92]		10	634	Electrolyte
Tokuda et al. [93]		10	712	Thermal fluids
Wu et al. [94]	$[C_6C_1im][NTf_2]$	10	745	Surfactant
Crosthwaite et al. [95]		10	700	High specific heat
	$[C_8C_1im][NTf_2]$	10	700	High specific heat
	$[C_4C_1pyr][NTf_2]$	10	670	High specific heat
Niu et al. [92]		10	625	Excellent ionic conductivity
Ji et al. [96]	$[C_4C_1Im][BF_4]$	10	588	CO_2 separation membrane
Knorr et al. [97]		10	>623	High thermal stability
Crosthwaite et al. [95]	$[C_6C_1im][BF_4]$	10	631	High specific heat
	$[C_8C_1im][BF_4]$	10	644	High specific heat
Nozari et al. [98]	$[C_4C_1im][PF_6]$	5	571	Gas separation
	$[C_4C_1C_1im][PF_6]$	5	615	Gas separation
Liaw et al. [99]	$[C_2C_1im][EtSO_4]$	10	>573	Thermal fluids
Gómez et al. [89]		–	575	High specific heat
Liaw et al. [99]	$[C_6C_1im][Cl]$	10	>473	Thermal fluids
Oliveira et al. [100]	$[C_2C_1im][C_2SO_3]$	10	899	High ionicity
	$[C_2C_1im][C_2SO_4]$	10	894	High ionicity

TABLE 4.9
Comparison of the Thermal Degradation (T_{onset}) for the Three Ionic Liquids at Different Heating Rates

	T_{onset}, K		
Heating Rate, K/min	$[C_2C_1im][Cl^a]$	$[C_4C_1pyr][Cl^a]$	$[C_4C_1pyr][NTf]^b$
5	529	513	885.5
10	536	526	913
15	542	536	920.5
20	555	545	935.6

[a] Williams et al. [101]
[b] Wang et al. [102]

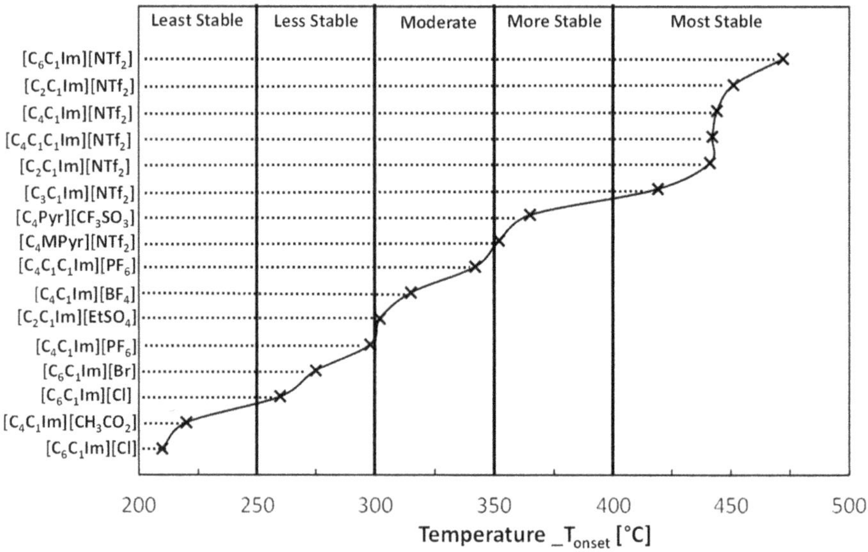

FIGURE 4.17 Thermal stability of common ionic liquids used as media for ionanocolloids.

who found that the open crucibles induce evaporation, whereas pierced and hermetical crucibles result in decomposition reactions [90]. Figure 4.18 illustrates the difference between conducting experiments with the crucible open or with a lid, and a small difference in the thermal degradation points is evident.

Despite the recognized importance of long-term thermal stability of ILs, relatively few studies have been published on this topic. Wu et al. [94] studied the flammability of $[C_6C_1Im][NTf_2]$ via an auto-ignition tester and TG/DSC analysis and

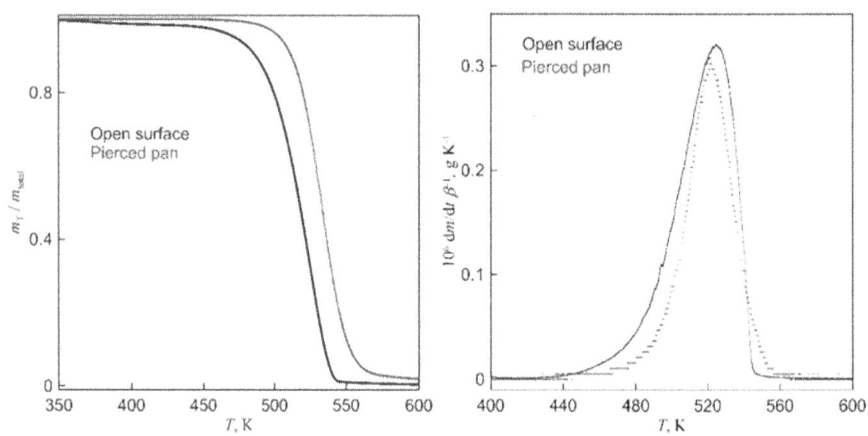

FIGURE 4.18 Comparison of data from open and pierced crucible for the heating rate of 5 K/min [91]. (Reprinted with permission from Elsevier.)

proved that AIT (auto-ignition temperature) is about 441.3°C. Seeberger et al. [107] investigated isothermal and non-isothermal TG stability of $[C_4C_1Im]$ $[NTf_2]$ at different heating rates, in the closed system, from 1 hour to 1 year. The experimental studies lead to a reliable estimation of the evaporation and decomposition rate with modeling of thermal degradation (see [107] for more details on ILs).

In conclusion, ILs exhibit exceptional thermal stability, often maintaining structural integrity well above 300 °C, depending on their ionic composition [11, 84, 108]. The thermal stability of ILs is a key factor influencing their performance in high-temperature applications such as ionanocolloids for heat transfer, energy storage, or solar thermal systems.

The purity and structural modification of the IL molecules subjected to changes in temperature are reflected in their melting and freezing behavior. DSC is a thermal analysis technique for the determination of phase transition temperatures and heat capacities, as well as the purity of materials. Melting point (T_m) is established as the onset of an endothermic peak, freezing temperature (T_f) as the onset of an exothermic peak (during the cooling cycle), cold crystallization temperature (T_{cc}) as the onset of an exothermic peak (during the heating cycle), and glass transition (T_g) as the midpoint of a small specific heat change (during the heating cycle). An evident illustration of the discrepancies that may occur in the reported values of the melting point data available is for example for the IL $[BMpyr][NTf_2]$: 266.5 K [109]; 263.84 K [110]; 262.2 K [111]; 261.4 K [112], 260.35 K [supplier Solvionic]; 258 K [113]; 253 K [114, suppliers including Iolitec and RocoGlobal]; 252 K [115]; and 250 K [116]. The influence of various factors is clearly reflected in these differences between reported values.

The chemical stability of neat ILs is relatively high, but when dissolved in aqueous solutions, their stability is strongly influenced by the nature of anions and cations. The hydrolytic stability of anions and cations is crucial for the pH stability of a mixture of ILs. For example, the imidazolium cation is more susceptible under basic conditions than the pyridinium or phosphonium cation [117], due to the high acidity at the C2 position. Cholinium-based ILs exhibit biodegradability and stability throughout a broad pH range; nevertheless, only two articles have been reported on the [Ch][Cl]/EG (choline chloride/ethylene glycol) mixture with MWCNTs [118] and $[Ch][NTf_2]$ (choline bis(trifluromethylsulfonyl)imide) with silver NPs [119]. At the same time, anions play an important role in the behavior of ILs with water. The hydrolytically anions are $[PF_6]$, $[BF_4]$, acetate, and carboxylic anions, but the most stable anion is $[NTf_2]$ with a wide range of pH stability between 2 and 12 [120].

A significant drawback in the application of ILs is their high cost, which poses a significant challenge for large-scale production. For instance, the current price for 100 g of pure IL such as $[C_4C_1im][NTf_2]$ is 380 euro, for $[C_4C_1im][BF_4]$ is 223 euro, for $[C_4C_1im][PF_6]$ is 324 euro, and for $[C_4C_1im][CF_3SO_3]$ 361 euro, which remains a price considerably high for scaled up production of ionanocolloids [https://www.sigmaaldrich.com]. As a result, developing new ionanocolloids at low cost and high quality is still a major challenge. However, recent studies have shown that ILs possess excellent thermal and chemical stability, allowing them to be recycled and making them promising candidates for nanofluids with long-term stability when incorporated appropriately [121].

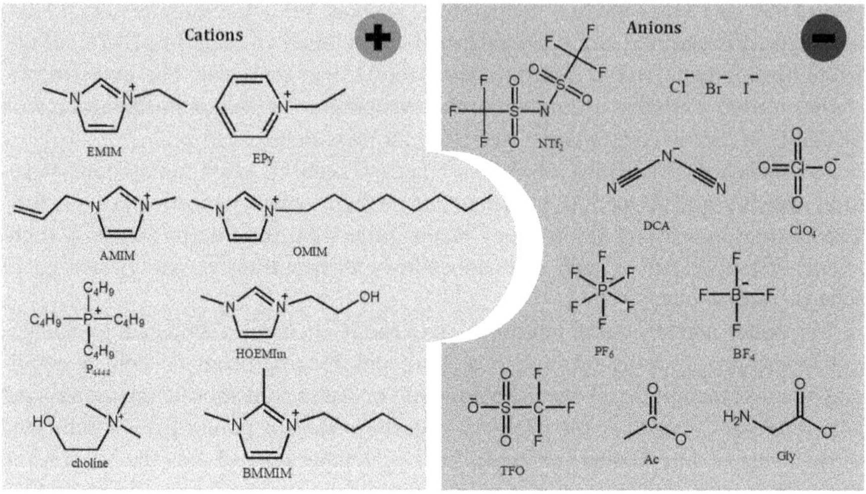

FIGURE 4.19 The common cations and anions present in ionic liquids used as media for nanoparticles.

The common cations and anions so far involved in ILs used as media for ionanocolloids are given below in Figure 4.19. Derivatives of imidazolium, pyridinium, and phosphonium are the common cations, and as anions the most used are NTf_2, BF_4, PF_6, and DCA anions.

4.4.2 Stability of Nanoparticles

NPs are materials with diameters between 1 and 100 nm, and according to the International Organization for Standardization (ISO), are defined as materials that exhibit external dimensions or internal surface structures at the nanoscale [122]. Structurally, nanomaterials are classified into four categories: 0D (metal and metal oxides), 1D (CNTs, carbon fibers, nanowires), 2D (graphene, GO, boron nitride (BN)), and 3D (bulk nanostructured materials of carbon), with possible morphologies as spheres, tubes, needles, sheets, and whiskers. These materials exhibit specific physical and chemical characteristics related to morphology, size, surface charge, concentration, and chemical stability, which provide extensive use in the preparation of ionanocolloids. NPs incorporated into ILs exhibit a wide range of applications, including use as refrigerants, in solar thermal systems, as materials with electrical insulation, and in enhancing the thermal and electrical conductivity of materials. The most selected NPs in ionanocolloids experiments [40], selected guided by the specific performance, are aluminum oxide (Al_2O_3), MWCNTs, graphene nanoplatelets (GNPs), graphene, graphite, mesoporous carbon (MC), zinc oxide (ZnO), iron oxide (Fe_2O_3), gold NPs (AuNP), silver NPs (AgNP), nanodiamond, BN, etc.

NP stability is extensively studied through properties such as aggregation behavior, composition, crystallinity, shape, size, and surface chemistry. To assess these

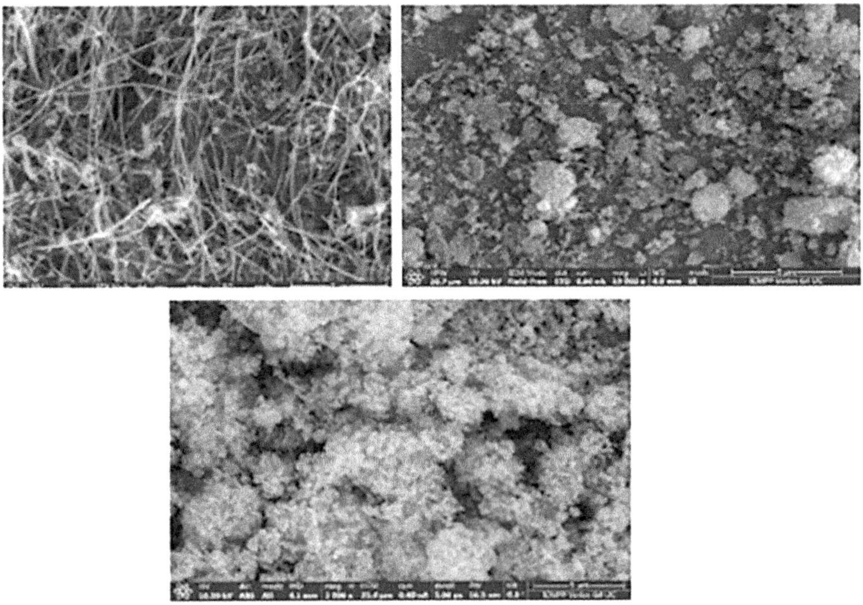

FIGURE 4.20 SEM images for NPs: first – MWCNT; second – Al_2O_3 and third – TiO_2 (at 5 μm scale).

characteristics, NPs are typically characterized using techniques such as X-ray diffraction (XRD), scanning electron microscopy (SEM), infrared spectroscopy (IR), and Raman spectroscopy, as well as TGA [123]. Numerous heat transfer fluids have been reported in the literature, in which NPs were characterized prior to their use in ionanocolloid experiments. For example, Chereches et al. [22, 124] examined by SEM the NPs as Al_2O_3, TiO_2, and MWCNT before investigating their isobaric specific heat properties, as represented in Figure 4.20.

The XRD patterns of pure NPs are also reported in order to determine their crystalline structure, phase purity, crystallite size, or degree of crystallinity. These structural parameters affect the performance of ionanocolloids in different applications, which is why the XRD is a widely used technique for NP characterization. For example, Worku et al. [125] for ZnO, Chereches et al. [124] for Al_2O_3, Chereches et al. [126] for MWCNT, and GNPs [127] obtain a morphological viewpoint of NPs used in the preparation of ionanocolloids.

After thermic stability of ILs, the thermogravimetric stability of NPs is an important parameter that indicates their thermal resistance and decomposition behavior under elevated temperatures. Commonly, the NPs used in ionanocolloids systems are highly temperature stable over 500°C, and as represented in Figure 4.21, the residual mass is not more than 20%. Diffractograms of MgO and TiO_2 reveal a small mass loss until 393 K that is due to the water moisture and depends on the crystallization forms, showing that the materials are stable over 973 K [67].

In conclusion, the stability of NPs is usually evaluated in order to establish the suitability of NPs for high-temperature applications in ionanocolloids [30].

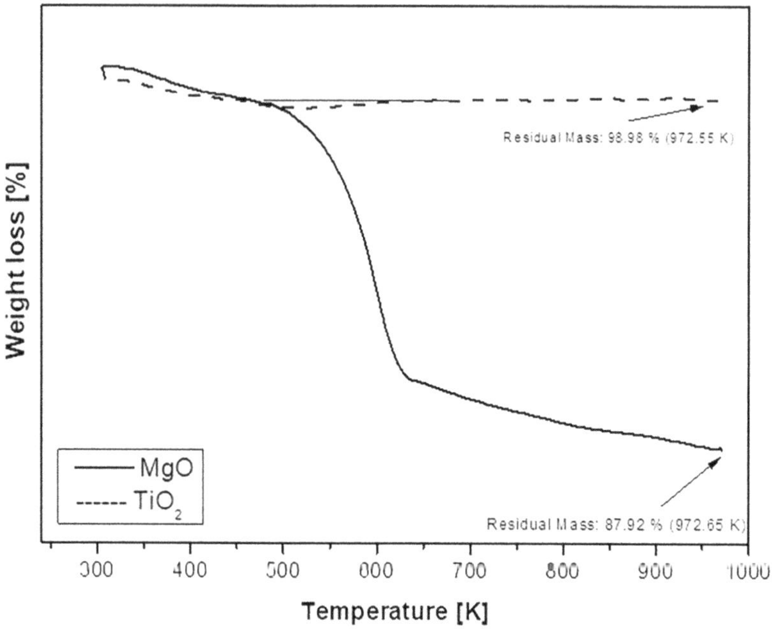

FIGURE 4.21 Thermogravimetric curves of MgO and TiO_2 nanoparticles used by Minea group (see more details on ionic liquids in [7, 22, 67, 124, 126]).

4.4.3 Ionanocolloids Stability

The concept of ionanocolloids was first introduced in 2009 by Nieto de Castro et al. [30], representing a novel class of heat transfer fluids that integrates various fields, including nanoscience, nanotechnology, thermofluids, and chemical and mechanical engineering. Ionanocolloids are defined as the dispersion of nanomaterials in various shapes, such as particles, tubes, or rods, within different ILs. Their long-term stability represents a major challenge to their industrial applications. Their stability is influenced by numerous aspects, and in order to understand and evaluate various techniques have been used. DLS (dynamic light scattering), Zeta potential (tunable resistive pulse sensing – TRPS), transmission electron microscopy (TEM), UV–VIS (ultraviolet–visible) spectrometry, sedimentation balance, and pH limits are always investigated and will be outlined below.

The agglomeration and sedimentation process of the NPs in the suspension of ionanocolloids play an important role in stability and subsequently the use as heat transfer agents. The stability of ionanocolloids appears to be associated with near neutral values of pH, because it was proven that very high or low values of pH may damage the heat transfer surface due to corrosion, especially at high temperature. As Chereches et al. [7] show, for example, for Al_2O_3 (weight fraction up to 0.5) dispersed in $[C_4C_1im][CH_3SO_3]$ as ILs, the stable system was determined at pH around 8.0–8.5, and no surfactant was needed to be added. Mukherjee and Paria [128], in a comprehensive review, show that the optimal pH values for stable aqueous dispersions vary by material; around 8 for alumina NPs, 9.5 for copper, and 2 for graphite. As expected, pH plays an important role in determining the zeta potential of suspension, which in turn governs their stability.

Generally, a suspension with an absolute zeta potential above 30 mV is considered moderately stable, while values exceeding 45 mV indicate good stability. Li et al. [129] demonstrated that coolant-based SiC nanofluids exhibit good stability over a 30-day period. Similar trends in both sedimentation behavior and zeta potential measurements were observed across different volume concentrations by Das et al. [130] for various formulated ILs + DEG/Al_2O_3 + CuO ionanocolloids (where DEG is diethylene glycol) up to 14 days. A clear illustration of the relationship between zeta potential, pH, and colloidal stability is presented in Figure 4.22. Anyway, most studies investigating the aspects such as pH interval, zeta potential value, and implicit stability are focused on nanofluids with low viscosity, particularly aqueous-based systems.

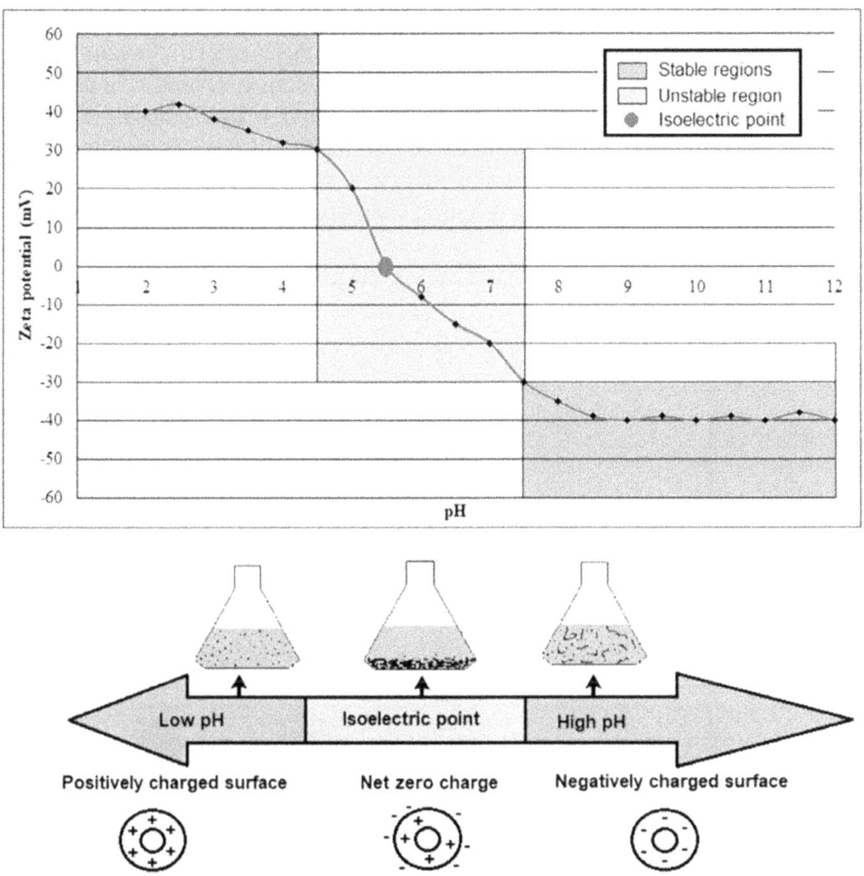

FIGURE 4.22 A general trend of zeta potential concerning pH and the influence on their stability [131]. (Reprinted with permission from Elsevier.)

Redel et al. [132] reported a stability of silver NPs in ILs for three days under an inert atmosphere. The authors proved, for the first time, a reproducible approach for the preparation of AgNPs in ILs. The precursors such as $AgBF_4$, $AgPF_6$, Ag_2O, or AgOTf were dissolved in different dried ILs ($[C_4C_1im][BF_4]$, $[C_4C_1im][PF_6]$, $[C_4C_1im]$ [OTf], $[C_4C_1im][NTf_2]$) and reacted with hydrogen (4 atm, 85°C) in the absence/presence of an n-butylimidazole as scavenger. Later, Corrêa et al. [133] prepared ionanocolloids based on Ag NPs, which exhibit stability exceeding one month. Following the same procedure as Redel et al. [132] described, AgNP_ $[C_4C_1im]$ $[NTf_2]$ with 10% of PVP (polyvinylpyrrolidone) has significant stability, evidenced by TEM images, which is further corroborated by the UV–VIS spectra [133].

GO dispersed in $[C_4C_1im][NTf_2]$ under ultrasonic agitation is stable over two weeks, as Hosseinghorbani et al. [73] reported in 2020. The dispersion was investigated using TEM and DLS. The TEM images of the 0.5 wt.% ionanocolloid show a well-dispersed GO with particles ranging from 20 to 70 nm over one week, and noticeable in aggregation with particle sizes shifting to the 70–150 nm range over the second week. Particle size measurements were also performed using DLS, and as detailed in Figure 4.23, the red curve indicates agglomeration over one week, since the blue curve indicates an increase in the particle diameter from 91.3 nm to 142 nm. Anyway, the visual inspection indicates no sedimentation or separation of phases.

Another effective method for characterizing nanofluids is **IR**, a valuable technique focused on (1) the structural characteristics and stability of ILs, (2) the interactions between NPs and ILs, and (3) the potential chemical modification of NP surfaces with ILs. NPs of Cu and Fe_3O_4, for instance, dispersed in various imidazolium-based

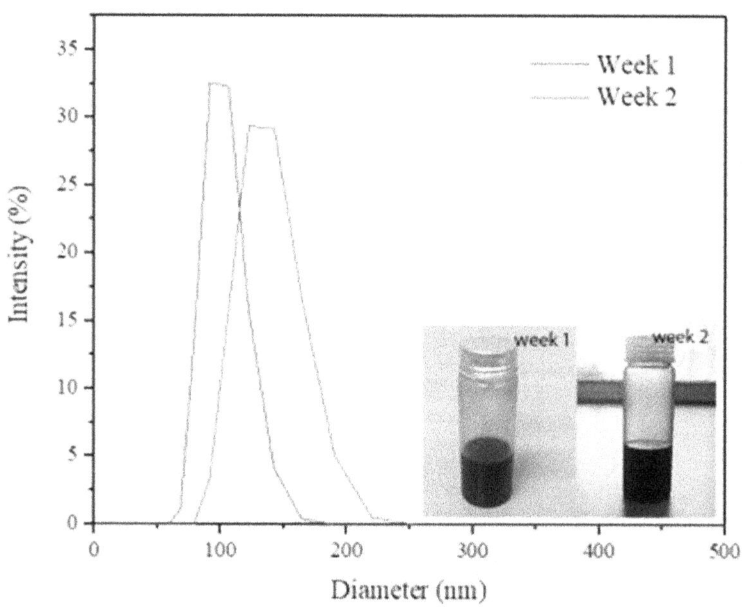

FIGURE 4.23 Nanoparticle size distribution of $C_4C_1imNTf_2$ containing 0.5 wt.% GO [73]. (Reprinted with permission from Elsevier.)

ILs ([C_2C_1Im][Ac], [C_2C_1Im][BF_4], [C_2C_1Im][$N(CN)_2$], [C_2C_1Im][OTf], [C_2C_1Im][NTf_2], [C_3C_1Im][BF_4], [C_4C_1Im][BF_4], [C_6C_1Im][BF_4] and [C_8C_1Im][BF_4] exhibited a notable alteration in fingerprint characteristics compared to pure IL [134]. The authors demonstrate that this phenomenon can be explained by the amplified scattering of generated light upon interaction with solid NPs, leading to an enhancement in absorption. The bands related to metal–ligand vibrations typically appear in the low frequency region, between 400 and 800 cm^{-1}, since for fluids, particularly for ILs, in the 3000–1000 cm^{-1}. However, it is important to note that the addition of NPs should not result in any chemical reaction with the molecules of the base fluid, and no significant differences should be observed in the IR spectra [135].

NMR (nuclear magnetic resonance) serves as a complementary method for identification of the stability of ILs in ionanocolloids. This technique is mostly for characterization of purity of ILs and for identification of potential reactions between NPs and ILs under varying time or temperature conditions [136].

The thermal decomposition profile of the mixtures of NPs with ILs proved an enhancement in thermal stability of the ionanocolloids [81], as presented in Table 4.10. It is obvious that the determined T_{onset} of ILs increases with the concentration of NPs, due to the higher stability of NPs compared to organic solvents (respectively, ILs). Babucci et al. [135] proved for a series of NPs, such as, for example, SiO_2, γ-Al_2O_3, and MgO immobilized in various ILs, that T_{onset} values decrease significantly. This could be motivated because the NPs based on metal oxide can act as catalysts for the decomposition of ILs at high temperature [135]. Also, a complete study by Oster et al. [81] indicates an approximately linear dependence of decomposition temperature (T_{onset}) with the concentration of NPs. Their main goal was to investigate IL-based nanofluids (ionanocolloids) for thermal applications using BN, CNTs, graphite, and MC NPs doped in five ILs [81].

As a conclusion, the available literature indicates a limited number of studies addressing the thermal stability of ionanocolloids formulated with various base ILs and NPs. Anyway, based on literature available, the thermal stability appears to be influenced by both the presence and type of NPs and thermal stability of ILs (see Table 4.10).

TABLE 4.10
Thermal Stability of Few Ionanocolloids: Influence of Ionic Liquid and Nanoparticle

Ionic Liquid	T_{onset} of Pure IL, K	T_{onset} of IL with MWCNT, K	T_{onset} of IL with 5 wt.% of xGnP, K	T_{onset} of IL with SiO_2, K	T_{onset} of IL with γ-Al_2O_3, K	T_{onset} of IL with MgO, K
[C_4C_1im][NTf_2]	717 [91] 695 [11]	–	–	633 [135]	569 [135]	479 [135]
[C_6C_1im][BF_4]	631 [95]	–	–	598 [135]	515 [135]	445 [135]
[$(C_6)_3PC_{14}$][NTf_2]	612 [137]	610 [137]	–	–	–	–
[C_2C_1py][$C_4F_9SO_3$]	685 [138]	–	688 [138]	–	–	–

4.5 CONCLUSION

Ionanocolloids are a new generation heat transfer fluid consisting of suspensions of ILs and NPs. Ionanocolloid/s is an inclusive expression by gathering the terms NP-enhanced IL, ionanofluids, and nanofluid under one roof with an approach that deals with the whole.

When ILs are used alone, their thermophysical properties are not satisfactory compared to conventional heat transfer fluids. However, adding NPs to ILs can improve their thermophysical properties more than adding NPs to conventional heat transfer fluids, and they can exhibit superior heat transfer properties.

In summary, several conclusions can be drawn as follows:

- The thermal conductivity of ILs is lower than that of regular heat transfer fluids, being comparable with that of the ILs.
- The NPs addition increases the thermal conductivity of ILs.
- The thermal conductivity variation with temperature of ionanocolloids follows the same trend as for ILs.
- The specific heat of ionanocolloids depends on some parameters such as NP type and morphology, concentration of NP, molecular and chemical structure of ILs, and temperature.
- In general, as the temperature increases, the specific heat value increases, and as the NP concentration increases, the specific heat value decreases.
- The density of ionanocolloids depends on some parameters such as NP type concentration and morphology, IL properties and temperature.
- Generally, the density decreases as the temperature increases, and the density increases as the NP concentration ratio increases.
- The thermal stability of ionanocolloids is strongly influenced by the structures of both the cation and the anion of ILs, with the anion typically having a larger effect on thermophysical characteristics.
- The NPs of any shape or structure have high thermal properties and dispersed in ILs generate thermally stable suspensions as ionanocolloids.
- Ionanocolloids exhibit high chemical stability at elevated temperatures, and their stability is significantly influenced by the pH of the medium.

ABBREVIATIONS

AIT	auto-ignition temperature
CNT	carbon nanotubes
DEG	diethylene glycol
DLS	dynamic light scattering
IL	ionic liquid
IR	infrared spectroscopy
MWCNT	multiwall carbon nanotubes
NMR	nuclear magnetic resonance
NP	nanoparticles
PVP	polyvinylpyrrolidone

SEM	scanning electron microscopy
SWCNT	single-wall carbon nanotubes
TEM	transmission electron microscopy
TGA	thermogravimetric analysis
TRPS	tunable resistive pulse sensing
UV–VIS	ultraviolet–visible
XRD	X-ray diffraction

ANNEX 1

Ionic Liquids – Abbreviations

[(C_6)3PC14)][NTf_2]	Trihexyltetradecylphosphoniumbis(trifluoromethylsulfo nyl)imide
[(C_6)3PC14)][Phosph]	Trihcxyltctradccylphosphoniumphosphinate
[Bmim][PF_6]	1-Butyl-3-methylimidazolium hexafluorophosphate
[C_2C1Im][C_2SO_4]	1-Ethyl-3-methylimidazolium ethylsulfate
[C_2mim][CH_3SO_3]	1-Ethyl-3-methylimidazolium methanesulfonate
[C_2mim][$EtSO_4$]	1-Ethyl-3-methyl imidazoliumethylsulfate
[C_4C1Im][Dca]	1-Butyl-3-methylimidazolium dicyanamide
[C_4C1Im][NTf_2]	1-Butyl-3-methylimidazolium bis(trifluoromethylsulfonyl)imide
[C_4C1Pyrr][NTf_2]	Butyl-1-methylpyrrolidinium bis(trifluoromethylsulfonyl) imide
[C_4mim][$(CF_3SO_2)_2N$]	1-n-Butyl-3-methylimidazolium bis(trifluoromethanesulf onylimide)
[C_4mim][BF_4]	1-Butyl-3-methylimidazolium tetrafluoroborate
[C_4mim][Br]	1-Butyl-3-methylimidazolium bromide
[C_4mim][CF_3SO_3]	1-n-Butyl-3-methylimidazoliumtrifluoromethanesulfonate
[C_4mim][Cl]	1-n-Butyl-3-methylimidazolium chloride
[C_4mim][I]	1-Butyl-3-methylimidazolium iodide
[C_4mim][$N(SO_2C_2F_5)_2$]	1-Butyl-3-methylimidazolium fluorinated sulfonyl imide
[C_4mim][NTf_2]	1-Butyl-3-methylimidazolium bis(trifluoromethylsulfonyl)imide
[C_4mim][PF_6]	1-Butyl-3-methylimidazolium hexafluorophosphate
[C_4mpyrr][NTf_2]	*N*-Butyl-*N*-methylpyrrolidiniumbis(trifluoromethanesulfo nyl)imide
[C_6C1Im][PF_6]	1-Hexyl-3-methylimidazolium hexafluorophosphate
[C_6mim][(CF3SO2)2N]	3-Hexyl-1-methylimidazolium bis(trifluoromethanesulfonyl)imide
[C_6mim][BF_4]	1-Hexyl-3-methylimidazolium tetrafluoroborate
[C6mim][NTf_2]	1-Hexyl-3-methylimidazolium bis(trifluoromethylsulfonyl)imide
[C6mim][PF_6]	1-Hexyl-3-methylimidazolium hexafluorophosphate
[C_8mim][(CF3SO2)2N]	3-Octyl-1-methylimidazolium bis(trifluoromethanesulfonyl)imide

[C8mim][NTf$_2$]	1-Methyl-3-octylimidazolium bis[trifluoromethylsulfonyl]imide
[Emim][DEP]	1-Ethyl-3-Methylimidazolium Diethyl Phosphate
[Emim][TFSI]	1-Ethyl-3-methylimidazolium bis(trifluoromethylsulfonyl)imide
[Hmim][BF$_4$]	1-Methylimidazolium tetrafluoroborate
[N$_{4111}$][NTf$_2$]	Butyltrimethylammoniumbis(trifluoromethylsulfonyl)imide
[(C$_1$)$_2$C$_2$HOC$_2$N][BF$_4$]	Ethyl-(2-hydroxyethyl)-dimethylammoniumtetrafluoroborate
[(C$_1$)$_2$C$_2$HOC$_2$N][PF$_6$]	Ethyl-(2-hydroxyethyl)-dimethylammoniumhexafluoroborate
[(C$_1$)$_2$C$_4$HOC$_2$N]Br	Butyl-(2-hydroxyethyl)-dimethylammoniumbromide
[(C$_1$)$_2$C$_6$HOC$_2$N][BF$_4$]	Hexyl(2-hydroxyethyl) dimethylammoniumtetrafluoroborate
[(C$_1$)$_2$C$_6$HOC$_2$N]Br	Hexyl(2-hydroxyethyl)dimethylammoniumbromide
[(C$_{10}$)$_2$(C$_1$)$_2$N][NO$_3$]	Didecyldimethylammonium nitrate
[(C$_6$OC)$_2$im][NTf$_2$]	1,3-Dihexyloxymethylimidazoliumbis(trifluoromethylsulfonyl)-amide
[1,3bmPY][CF$_3$SO$_3$]	1-Butyl-3-methylpyridiniumtrifluoromethanesulfonate
[1,4bmPY][NTf$_2$]	1-Butyl-4-methylpyridiniumbis(trifluoromethylsulfonyl)-amide
[1,4bmPY][TOS]	1-Butyl-4-methylpyridinium tosylate
[4-C$_4$py][BF$_4$]	4-Butyl-pyridinium tetrafluoroborate
[Be(C$_1$)C$_n$N][NO$_3$]	(Benzyl)dimethylalkylammonium nitrate
[bmim][BF$_4$]	1-Butyl-3-methylimidazolium tetrafluoroborate
[bmim][CF$_3$SO$_3$]	1-Butyl-3-methylimidazoliumtrifluoromethanesulfonate
[bmim][DCA]	1-Butyl-3-methylimidazolium dicyanamide
[bmim][MDEGSO$_4$]	1-Butyl-3-methylimidazolium2-(2-methoxyethoxy) ethyl sulfate
[bmim][NO$_3$]	1-Butyl-3-methylimidazolium nitrate
[bmim][OcSO$_4$]	1-Butyl-3-methylimidazolium octylsulfate
[bmim][PF$_6$]	1-Butyl-3-methylimidazolium hexafluorophosphate
[bmim][SCN]	1-Butyl-3-methylimidazolium thiocynate
[bmim][TfO]	1-Butyl-3-methylimidazolium trifluoromethanesulfonate
[BMP][Tf$_2$N]	1-Butyl-1-methylpyrrolidinium bis(trifluoromethylsulfonyl)imide
[bmPYR][CF$_3$SO$_3$]	1-Butyl-1-methylpyrrolidiniumtrifluoromethanesulfonate
[(C$_{10}$H$_{21}$OCH$_2$)$_2$Im][Tf$_2$N]	1,3-Didecyloxymethylimidazolium*bis*(trifluoro-methylsulfonyl)imide
[C$_1$C$_1$Im][C$_1$SO$_4$]	1,3-Dimethylimidazolium methylsulfate
[C$_2$C$_1$Im][C$_2$SO$_4$]	1-Ethyl-3-methylimidazolium ethylsulfate
[C$_2$C$_1$Im][PF$_6$]	1-Ethyl-3-methylimidazoliumhexafluorophosphate

[C$_2$C$_1$Im][TOS]	1-Ethyl-3-methylimidazolium tosylate
[C$_4$C$_1$Im][C$_1$SO$_4$]	1-Butyl-3-methylimidazoliummethylsulfate
[C$_4$C$_1$Im][C$_8$SO$_4$]	1-Butyl-3-methylimidazolium octylsulfate
[C$_4$C$_1$Im][MDEGSO$_4$]	1-Butyl-3-methylimidazolium2-(2-methoxyethoxy)-ethysulfate
[C$_4$C$_1$Im][PF$_6$]	1-Butyl-3-methylimidazoliumhexafluorophosphate
[C$_4$C$_1$Im]Cl	1-Butyl-3-methylimidazolium chloride
[C4H9OCH2–C1Im][BF4]	1,3-Dibutyloxymethylimidazoliumtetrafluoroborate
[C6C1Im][Tf2N]	1-Hexyl-3-methylimidazolium*bis*(trifluoromethylsulfonyl)imide
[C6H13OCH2–C1Im][BF4]	1-Hexyloxymethyl-3-methylimidazoliumtetrafluoroborate
[C6H13OCH2–C1Im][Tf2N]	1-Hexyloxymethyl-3-methylimidazolium*bis*(trifluoromethylsulfonyl)imide
[C$_6$mim][BF$_4$]	1-Hcxyl-3-mcthylimidazolium tetrafluoroborate
[C$_6$mim][PF$_6$]	1-Hexyl-3-methylimidazolium hexafluorophosphate
[C$_6$OCmim][NTf$_2$]	1-Hexyloxymethyl-3-methyl-imidazolium bis(trifluoromethylsulfonyl)-amide
[C$_8$H$_{17}$OCH$_2$)$_2$Im][Tf$_2$N]	1,3-Dioctyloxymethylimidazolium*bis*(trifluoromethylsulfonyl)imide
[C$_8$mim][BF$_4$]	1-Octyl-3-methylimidazolium tetrafluoroborate
[C$_8$mim][PF$_6$]	1-Octyl-3-methylimidazolium hexafluorophosphate
[C$_9$mim][PF$_6$]	1-Nonyl-3-methylimidazolium hexafluorophosphate
[emim][Ac]	1-Ethyl-3-methylimidazolium acetate
[emim][BF$_4$]	1-Ethyl-3-methylimidazolium tetrafluoroborate
[emim][C$_2$N$_3$]	1-Ethyl-3-methylimidazolium dicyanamide
[emim][EtSO$_4$]	1-Ethyl-3-methylimidazolium ethylsulfate
[emim][Tf$_2$N]	1-Ethyl-3-methylimidazolium bis(trifluoromethylsulfonyl)imide
[emim][TFA]	1-Ethyl-3-methylimidazolium trifluoroacetate
[Et$_3$S][NTf$_3$]	Triethyl-sulfoniumbis(trifluoromethylsulfonyl)-amide
[Hmim][NTf$_2$]	1-Hexyl-3-methylimidazoliumbis(trifluoromethylsulfonyl)-amide
[N-bupy][BF$_4$]	N-Butylpyridinium tetrafluoroborate
[NH$_2$p-bim][BF$_4$]	1-Propylamide-3-butyl imidazolium tetrafluoroborate
[P14,6,6,6][Tf$_2$N]	Trihexyltetradecylphosphonium bis(trifluoromethylsulfonyl)imide
HEAA	2-(2-Hydroxy ethoxy)-ammonium acetate
HEAF	2-(2-Hydroxy ethoxy)-ammonium formate
HEF	2-Hydroxy ethylammonium formate
THEAA	Tri-(2-hydroxy ethyl)-ammonium acetate
[C$_4$mim]Br	3-Butyl-1-methylimidazolium bromide
[C$_8$mim]Br	3-Octyl-1-methylimidazolium bromide
[C$_8$mim]Br	3-Octyl-1-methylimidazolium bromide

[C_{10}mim]Br	3-Decyl-1-methylimidazolium bromide
[C_4dmim]Br	3-Butyl-1,2-dimethylimidazolium bromide
[C_6dmim]Br	3-Hexyl-1,2-dimethylimidazolium bromide
[C_8dmim]Br	3-Octyl-1,2-dimethylimidazolium bromide
[C_{10}dmim]Br	3-Decyl-1,2-dimethylimidazolium bromide
[C_4mpyrrol]Br	N,N-Butylmethylpyrrolidinium bromide
[C_8mpyrrol]Br	N,N-Octylmethylpyrrolidinium bromide
[C_8mpyrrol]Br	N,N-Octylmethylpyrrolidinium bromide
[C_4mim][$(CF_3SO_2)_2N$]	3-Butyl-1-methylimidazolium bis(trifluoromethanesulfonyl)imide
[C_6mim][$(CF_3SO_2)_2N$]	3-Hexyl-1-methylimidazolium bis(trifluoromethanesulfonyl)imide
[C_8mim][$(CF_3SO_2)_2N$]	3-Octyl-1-methylimidazolium bis(trifluoromethanesulfonyl)imide
[C_{10}mim][$(CF_3SO_2)_2N$]	3-Decyl-1-methylimidazolium bis(trifluoromethanesulfonyl)imide
[C_2dmim][$(CF_3SO_2)_2N$]	3-Ethyl-1, 2-dimethylimidazolium bis(trifluoromethanesulfonyl)imide
[C_4dmim][$(CF_3SO_2)_2N$]	3-Butyl-1, 2-dimethylimidazolium bis(trifluoromethanesulfonyl)imide
[C_6dmim][$(CF_3SO_2)_2N$]	3-Hexyl-1, 2-dimethylimidazolium bis(trifluoromethanesulfonyl)imide
[C_8dmim][$(CF_3SO_2)_2N$]	3-Octyl-1, 2-dimethylimidazolium bis(trifluoromethanesulfonyl)imide
[C_{10}dmim][$(CF_3SO_2)_2N$]	3-Decyl-1, 2-dimethylimidazolium bis(trifluoromethanesulfonyl)imide
[C_4mpyrrol] [$(CF_3SO_2)_2N$]	N,N-Butylmethylpyrrolidiniumbis(trifluoromethanesulfonyl)imide
[C_8mpyrrol][$(CF_3SO_2)_2N$]	N,N-Octylmethylpyrrolidiniumbis(trifluoromethanesulfonyl)imide
[P_{66614}][$(CF_3SO_2)_2N$]	Trihexyltetradecylphosphoniumbis(trifluoromethanesulfonyl)imide
[$C_6H_{13}OCOCH_2$dmim]Br	3-Hexoxycarbonylmethyl-1,2-dimethylimidazolium bromide
[$C_6H_{13}OCOCH_2$mpyrrol]Br	*N*-Hexoxycarbonylmethyl-N-methylpyrrolidinium bromide
[$C_8H_{17}OCOCH_2$dmim]Br	3-Octoxycarbonylmethyl-1,2-dimethylimidazolium bromide

REFERENCES

1. A.A. Minea, Overview of Ionic Liquids as Candidates for New Heat Transfer Fluids, International Journal of Thermophysics 41 (2020) 151.
2. Therminol 68. Technical Bulletin. Louvain-la-Neuve, Belgium: Solutia Europe, 2014.
3. Y.A. Sanmamed, D. Gonzalez-Salgado, J. Troncoso, L. Romani, A. Baylaucq, C. Boned, Experimental Methodology for Precise Determination of Density of RTIonic Liquids as a Function of Temperature and Pressure Using Vibrating Tube Densimeters, The Journal of Chemical Thermodynamics 42(4) (2010) 553–563.

4. S. Kakinuma, T. Ishida, H. Shirota, Femtosecond Raman-Induced Kerr Effect Study of Temperature-Dependent Intermolecular Dynamics in Imidazolium-Based Ionic Liquids: Effects of Anion Species and Cation Alkyl Groups, The Journal of Physical Chemistry B 121(1) (2017) 250–264.
5. V. V. Wadekar, Ionic Liquids as Heat Transfer Fluids – An Assessment Using Industrial Exchanger Geometries, Applied Thermal Engineering 111 (2017) 1581–1587.
6. F. Wang, L. Han, Z. Zhang, X. Fang, J. Shi, W. M,. Surfactant-Free Ionic Liquid-Based Nanofluids With Remarkable Thermal Conductivity Enhancement at Very Low Loading of Graphene. Nanoscale Research Letters 7 (2012) 314–319.
7. E.I. Cherecheş, J.I. Prado, M. Cherecheş, A.A. Minea, L. Lugo, Experimental Study on Thermophysical Properties of Alumina Nanoparticle Enhanced Ionic Liquids, Journal of Molecular Liquids 291 (2019) 111332.
8. T.C. Paul, Investigation of thermal performance of nanoparticle enhanced ionic liquids (NEIonic liquids) for solar collector applications. (Doctoral dissertation). (2014). Retrieved from http://scholarcommons.sc.edu/etd/2873.
9. J.M.P. França, S.I.C. Vieira, M.J.V. Lourenço, S.M.S. Murshed, C.A. Nieto de Castro, Thermal Conductivity of Ionanocolloids of [C4mim][NTf2] and [C2mim][EtSO4] with Carbon Nanotubes. Experiment and Theory, Journal of Chemical and Engineering Data 58 (2013) 467–476.
10. V. Patil, A. Cera-Manjarre, D. Salavera, C. Rode, K. Patil, C.A. Nieto de Castro, A. Coronas, Ru-Imidazolium Halide Ionanocolloids. Synthesis, Structural, Morphological and Thermophysical Properties, Journal of Nanofluids 5 (2016) 191–208.
11. C.P. Fredlake, J.M. Crosthwaite, D.G. Hert, S.N.V.K. Aki, J.F. Brennecke, Thermophysical Properties of Imidazolium-Based Ionic Liquids. Journal of Chemical & Engineering Data 49(4), (2004) 954–964.
12. O. Yamamuro, M. Kofu, Calorimetric and Neutron Scattering Studies on Glass Transitions and Ionic Diffusions in Imidazolium-Based Ionic Liquids, IOP Conference Series: Materials Science and Engineering 196, (2017) 012001-1–012001–4.
13. P.C. Ribeiro, S.I.C. Vieira, P. Goodrich, C. Hardacre, M.J. V. Lourenço, C.A. Nieto de Castro, Thermal Conductivity of [Cnmim] [(CF3SO2)2N] and [C4mim][BF4] Ionanocolloids with Carbon Nanotubes – Measurement, Theory and Structural Characterization, Journal of Nanofluids 2 (2013) 55–62.
14. Y.U. Paulechka, A.V. Blokhin, G.J. Kabo, Evaluation of Thermodynamic Properties for Non-Crystallizable Ionic Liquids, Thermochimica Acta 604, (2015) 122–128.
15. A.A. Minea, S.M.S. Murshed, Ionic Liquids-Based Nanocolloids—A Review of Progress and Prospects in Convective Heat Transfer Applications, Nanomaterials 11(4) (2021), 1039.
16. E.I. Cherecheş, D. Bejan, C. Ibanescu, M. Danu, A.A. Minea, Ionanocolloids with [C2mim][CH3SO3] Ionic Liquid and Alumina Nanoparticles: An Experimental Study on Viscosity, Specific Heat and Electrical Conductivity, Chemical Engineering Science 229 (2021) 116140.
17. E.I. Cherecheş, A.A. Minea, Experimental Evaluation of Electrical Conductivity of Ionanocolloids Based on Water – [C2mim][CH3SO3] Ionic Liquids Mixtures and Alumina Nanoparticles, Journal of Thermal Analysis and Calorimetry 145 (6) (2021) 3151–3157.
18. E.I. Cherecheş, J.I. Prado, C. Ibanescu, M. Danu, A.A. Minea, L. Lugo, Viscosity and Isobaric Specific Heat of Alumina Nanoparticle Enhanced Ionic Liquids: an Experimental Approach, Journal of Molecular Liquids, (2020) 114020.
19. E.I. Cherecheş, M. Cherecheş, A. Alexandru, A. Dima, A.A. Minea, Nanoparticles in Ionic Liquids: Numerical Evaluation of Heat Transfer Behaviour in Laminar Flow, Heat Transfer Engineering 42 (2020) 1818372.
20. A.A. Minea, S.M.S. Murshed, A Review on Development of Ionic Liquid Based Nanofluids and Their Heat Transfer Behavior, Renewable and Sustainable Energy Reviews, 91 (2018) 584–599.

21. A.A. Minea, E.I. Cherecheș, Experimental Studies on Thermal Conductivity and Heat Transfer of 1-Butyl-3-Methylimidazolium Tetrafluoroborate Ionic Liquid and Its Nanocolloids, International Communications in Heat and Mass Transfer 154 (2024) 107406.
22. E.I. Cherecheș, D. Bejan, A.A. Minea, Experimental on Viscosity and Isobaric Specific Heat of [C4mim][BF4] Ionic Liquid with MWCNT Nanoparticles, Journal of Molecular Liquids 391 (2023) 123214.
23. E.I. Cherecheș, D. Bejan, C. Ibanescu, M. Danu, A.A. Minea, Experimental Investigation of Isobaric Specific Heat and Viscosity for Suspensions of Alumina Nanoparticles in [C4mim][BF4] Ionic Liquid, Journal of Thermal Analysis and Calorimetry 148 (2023) 8879–8888.
24. E.I. Cherecheș, C. Ibanescu, M. Danu, A.A. Minea, Studies on Rheological Properties and Isobaric Specific Heat of ZnO - [C4mim][BF4] Nanoparticle Enhanced Ionic Liquid, Journal of Molecular Liquids 380 (2023) 121759.
25. A.A. Minea, E.I. Cherecheș, State of the Art on the Development of 1-Butyl-3-Methylimidazolium Tetrafluoroborate Ionic Liquid and Its Nanocolloids, Journal of Molecular Liquids 368 (2022) 120691.
26. C.A. Nieto de Castro, S.M.S. Murshed, M.J.V. Lourenço, F.J.V. Santos, M.L.M Lopes, J.M.P. França, Enhanced Thermal Conductivity and Specific Heat of Carbon Nanotubes-Ionanocolloids, International Journal of Thermal Sciences 62 (2012) 34–39.
27. A.P.C. Ribeiro, S.I.C. Vieira, J.M.P. França, C.S. Queirós, E. Langa, M.J.V. Lourenço, S.M.S. Murshed, C.A. Nieto de Castro, Thermal Properties of Ionic Liquids and Ionanocolloids. In: A. Kokorin (ed.) Ionic Liquids: Theory, Properties, New Approaches. Rijeka: InTech, 2011.
28. S. Aman, I. Khan, Z. Ismail, M. Z. Salleh, Q.M. Al-Mdallal, Heat Transfer Enhancement in Free Convection Flow of CNTs Maxwell Nanofluids with Four Different Types of Molecular Liquids. Scientific Reports 7 (2017) 2445.
29. D. Tomida, S. Kenmochi, T. Tsukada, K. Qiao, Q. Bao, C. Yokoyama, Viscosity and Thermal Conductivity of 1-Hexyl-3-Methylimidazolium Tetrafluoroborate and 1-Octyl-3-Methylimidazolium Tetrafluoroborate at Pressures up to 20 MPa, International Journal of Thermophysic 33 (2012) 959–969.
30. C.A. Nieto de Castro, M.J.V. Lourenço, A.P.C. Ribeiro, E. Langa, S.I.C. Vieira, P. Goodrich, C. Hardacre, Thermal Properties of Ionic Liquids and Ionanocolloids of Imidazolium and Pyrrolidinium Liquids, Journal of Chemical and Engineering Data 55 (2010) 65–69.
31. V.M.B. Nunes, M.J.V. Lourenço, F.J.V. Santos, M.M.S. Matos Lopes, C.A. Nieto de Castro, Accurate Measurements of Physico-Chemical Properties on Ionic Liquids and Molten Salts, In: K. E. Seddon and M. Gaune-Escard (eds.), Ionic Liquids and Molten Salts: Never the Twain. London: John Wiley, 2010.
32. R. Ge, C. Hardacre, P. Nancarrow, D.W. Rooney, Thermal Conductivities of Ionic Liquids Over the Temperature Range from 293 K to 353 K, Journal of Chemical and Engineering Data 52 (2007) 1819–1825.
33. S.M.S. Murshed, K.C. Leong, C. Yang, Thermophysical and Electrokinetic Properties of Nanofluids – A Critical Review, Applied Thermal Engineering 28 (2008) 2109–2125.
34. S.V. Dzyuba, R.A. Bartsch, Influence of Structural Variations in 1-Alkyl(aralkyl)-3-Methylimidazolium Hexafluorophosphates and Bis(trifluoromethyl-Sulfonyl)imides on Physical Properties of the Ionic Liquids, ChemPhysChem 3 (2002) 161–166.
35. S.M.S. Murshed, P. Estellé, A State of the Art Review on Viscosity of Nanofluids, Renewable and Sustainable Energy Reviews 76 (2017) 1134–1152.
36. E.B. Fox, A.E. Visser, N.J. Bridges, J.W. Amoroso, Thermophysical Properties of Nanoparticle-Enhanced Ionic Liquids (NEIonic Liquids) Heat-Transfer Fluids, Energy Fuels 27 (2013) 3385–3393.

37. C.A. Nieto de Castro, S.M.S. Murshed, M.J.V. Lourenço, F.J.V. Santos, M.L. Matos Lopes, J.M.P. França, Ionanocolloids: New Heat Transfer Fluids for Green Processes Development, Green Solvents I: Properties and Applications in Chemistry. Netherlands: Springer, 2012.
38. S.M.S. Murshed, C.A. Nieto de Castro, M.J.V. Lourenço, J. França, A.P.C. Ribeiro, S.I.C. Vieira, C.S. Queirós, Ionanocolloids as Novel Fluids for Advanced Heat Transfer Applications, World Academy of Science, Engineering and Technology 52 (2011) 795–798.
39. L.S. Sundar, Ionic-Liquid-Based Nanofluids and Their Heat-Transfer Applications: A Comprehensive Review. ChemPhysChem 1–37 (2023) 202300191.
40. L.S. Sundar, F. Shaik, Z. Said, Experimental Determination of Thermophysical Properties and Figures-of-Merit Analysis of 80:20% Water and Ionic Liquid Mixture Based MXene Nanofluid. Proceedings of the Institution of Mechanical Engineers, Part C: Journal of Mechanical Engineering Science 237 (17) (2023) 3916–3930.
41. L.S. Sundar, S.M.O. Tavares, E. V. Ramana, A.M.B. Pereira, Thermophysical Properties, and Figures-of-Merit Analyses of nanodiamond/CuO Ionanofluids: Experimental and Artificial Neural Network Predictions, Journal of Ionic Liquids 4 (2024) 100113.
42. L.S. Sundar, Graphene Oxide Based [EMIM]Cl Ionanofluids in a Tube and Their Heat Transfer, and Friction Factor Analyses Under High Prandtl Numbers: Experimental and ANN Predictions, Journal of Ionic Liquids 5 (2025) 100144.
43. M. Moosavi, M. Torkzadeh, Z. Akbarinezhad, Molecular Dynamics Investigation of Ionanofluids (INFs): Towards a Deeper Understanding of Their Thermophysical, Structural and Dynamical Properties, Journal of Molecular Liquids 399 (2024) 124355
44. R. Jędrysiak, A. Kolanowska, A. Blacha, S. Boncel, M. Dzida, Water-Like Thermal Conductivity of Ionanofluids Containing High Aspect Ratio Multi-Walled Carbon Nanotubes and 1-Ethyl-3-Methylimidazolium-Based Ionic Liquids with Cyanofunctionalized Anions, Journal of Molecular Liquids 391 (2023) 123329
45. T. Yasmin, S.O. Giwa, S. Noor, M. Sharifpur, Thermal Conductivity Enhancement of Metal Oxide Nanofluids: A Critical Review, Nanomaterials 13 (2023) 597.
46. H. Xie, Z. Zhao, J. Zhao, H. Gao, Measurement of Thermal Conductivity, Viscosity and Density of Ionic Liquid [EM-IM][DEP]-Based Nanofluids, Chinese Journal of Chemical Engineering 24 (2016) 331–338.
47. E. Fabre, S.M.S. Murshed, A Comprehensive Review of Thermophysical Properties and Prospects of Ionanocolloids in Thermal Energy Applications, Renewable and Sustainable Energy Reviews 151 (2021) 111593.
48. J. Pinto, E. Fabre, S.M.S. Murshed, Evaluation of Stability, Viscosity and Electrical Conductivity of [C2mim][DCA] Based Ionanocolloids, Heat and Mass Transfer 60 (2024) 2073–2079.
49. S. Faraji, H. Shekaari, M.T. Zafarani-Moattar, M. Mokhtarpour, Experimental Studies on Thermophysical Properties of Protic Ionic Liquids for Thermal Energy Storage Systems, Journal of Energy Storage 54 (2022) 105251.
50. M. Kraiem, M. Karkri, M. Fois, P. Sobolciak, Thermophysical Characterization of Paraffins Versus Temperature for Thermal Energy Storage, Buildings 13 (2023).
51. W. Dong, H. Tian, W. Zhang, J.-J. Zhou, X. Pang, Development of NaCl–MgCl2–CaCl2 Ternary Salt for High-Temperature Thermal Energy Storage Using Machine Learning, ACS Applied Materials & Interfaces 16 (2024) 530–539.
52. Z. Rong, G. Pan, J. Lu, S. Liu, J. Ding, W. Wang, D.-J. Lee, Ab-Initio Molecular Dynamics Study on Thermal Property of NaCl–CaCl2 Molten Salt for High-Temperature Heat Transfer and Storage, Renewable Energy 163 (2021) 579–588.
53. X. Yu, J. Chang, R. Huang, Y. Huang, Y. Lu, Z. Li, L. Wang, Sensitivity Analysis of Thermophysical Properties on PCM Selection Under Steady and Fluctuating Heat Sources: A Comparative Study, Applied Thermal Engineering 186 (2021) 116527.

54. R. Yang, L. Yuan, D. Zhang, T. Wu, Y. Lu, Dynamic Heat Transfer Mechanisms of Internal Thermal Mass: Effects of Thermal Conductivity and Diffusivity Under Varied Temperature Conditions, Case Studies in Thermal Engineering 65 (2025) 105600.
55. M. Shamsaei, A. Carter, M. Vaillancourt, A Review on the Heat Transfer in Asphalt Pavements and Urban Heat Island Mitigation Methods, Construction and Building Materials 359 (2022) 129350.
56. Z. Zhang, Y. Wang, W. Yao, F. Gao, C. Shou, Effect of Thermo-Physical Parameters on Heat Transfer Characteristics of the Wall Implanted with Heat Pipes, Applied Thermal Engineering 210 (2022) 118375.
57. K. Elsaid, M.A. Abdelkareem, H.M. Maghrabie, E.T. Sayed, T. Wilberforce, A. Baroutaji, A.G. Olabi, Thermophysical Properties of Graphene-Based Nanofluids, International Journal of Thermofluids 10 (2021) 100073.
58. H. Hasen, B. Abdulmajeed, Theoretical Investigation of the Density and the Specific Heat of [EMIM][BF4] and Its MWCNTs Ionanocolloids: Effect of Temperature and MWCNTs Concentration. Journal of Physics: Conference Series (2021) 2114.
59. K. Oster, C. Hardacre, J. Jacquemin, A. Ribeiro, A. El-Sinawi, Understanding the Specific Heat Enhancement in Ionic Liquid-Based Nanofluids (ionanocolloids). Journal of Molecular Liquids, 253 (2018) 326–339.
60. P.K. Kanti, E.I. Chereches, A.A. Minea, K. V Sharma, Experiments on Thermal Properties of Ionic Liquid Enhanced with Alumina Nanoparticles for Solar Applications, Journal of Thermal Analysis and Calorimetry 147 (2022) 13027–13038.
61. M. Lasfargues, Q. Geng, H. Cao, Y. Ding, Mechanical Dispersion of Nanoparticles and Its Effect on the Specific Heat of Impure Binary Nitrate Salt Mixtures, Nanomaterials 5 (2015) 1136–1146.
62. R. Hentschke, On the Specific Heat Enhancement in Nanofluids, Nanoscale Research Letters 11 (2016) 88.
63. H. Riazi, T. Murphy, G.B. Webber, R. Atkin, S.S.M. Tehrani, R.A. Taylor, Specific Heat Control of Nanofluids: A Critical Review, International Journal of Thermal Sciences 107 (2016) 25–38.
64. L. Das, F. Rubbi, K. Habib, N. Aslfattahi, R. Saidur, B. Baran Saha, S. Algarni, K. Irshad, T. Alqahtani, State-of-the-Art Ionic Liquid & Ionanocolloids Incorporated with Advanced Nanomaterials for Solar Energy Applications, Journal of Molecular Liquids 336 (2021) 116563.
65. N. Parmar, M. Bendová, Z. Wagner, J. Jacquemin, A Study of Changes in the Specific Heat of Carbon Nanotube-Based Ionanocolloids Prepared from a Series of Imidazolium Ionic Liquids, Physical Chemistry Chemical Physics 24 (2022) 22181–22190.
66. B. Jóźwiak, G. Dzido, E. Zorębski, A. Kolanowska, R. Jędrysiak, J. Dziadosz, M. Libera, S. Boncel, M. Dzida, Remarkable Thermal Conductivity Enhancement in Carbon-Based Ionanocolloids: Effect of Nanoparticle Morphology, ACS Applied Materials & Interfaces 12 (2020) 38113–38123.
67. E.I. Chereches, D. Bejan, C. Ibanescu, M. Danu, A.A. Minea, Nanocolloids Based on PEG 400 with MgO Nanoparticles: Experimental Study on Viscosity and Specific Heat, Thermal Science and Engineering Progress 43 (2023) 101985.
68. B. Wang, L. Lin, S. Ren, W. Wu, Specific Heat of Non-Functional and Functional Ionic Liquids During the Absorption of SO2, Industrial & Engineering Chemistry Research 60 (2021) 13740–13747.
69. P. Kanti, A.A. Minea, K. V Sharma, M. Revanasiddappa, Improved Thermophysical Properties of Graphene Ionanocolloid as Heat Transfer Fluids for Thermal Applications, Journal of Ionic Liquids 2 (2022) 100038.
70. L.S. Sundar, S.M.O. Tavares, E.V. Ramana, A.M.B. Pereira, Thermophysical Properties, and Figures-of-Merit Analyses of Nanodiamond/CuO Ionanocolloids: Experimental and Artificial Neural Network Predictions, Journal of Ionic Liquids 4 (2024) 100113.

71. G. Huminic, A. Huminic, Heat Transfer Capability of Ionanocolloids for Heat Transfer Applications, International Journal of Thermophysics 42 (2020) 12.
72. K. Cwynar, J. Dziadosz, Ł. Scheller, E. Zorębski, R. Jędrysiak, A. Kolanowska, A. Blacha, S. Boncel, M. Dzida, On Isobaric Specific Heat of Ionanocolloids with Carbon Nanotubes – An Experimental Study, Journal of Molecular Liquids 387 (2023) 122535.
73. A. Hosseinghorbani, M. Mozaffarian, G. Pazuki, Application of Graphene Oxide Ionanocolloid as a Superior Heat Transfer Fluid in Concentrated Solar Power Plants, International Communications in Heat and Mass Transfer 111 (2020) 104450.
74. S.Q. Zhou, R. Ni, Measurement of the Specific Heat of Water-Based Al2O3 Nanofluid, Applied Physics Letters 92 (2008) 93123.
75. J. Liu, F. Wang, L. Zhang, X. Fang, Z. Zhang, Thermodynamic Properties and Thermal Stability of Ionic Liquid-Based Nanofluids Containing Graphene as Advanced Heat Transfer Fluids for Medium-to-High-Temperature Applications, Renewable Energy 63 (2014) 519–523.
76. T. Myrdek, C. Popescu, W. Kunz, Physical-Chemical Properties of Newly Synthesized Tetraalkylammonium Alkyl Ether Carboxylate Ionic Liquids, Journal of Molecular Liquids 322 (2021) 114947.
77. K. Paduszyński, Extensive Databases and Group Contribution QSPRs of Ionic Liquids Properties. 1. Density, Industrial & Engineering Chemistry Research 58 (2019) 5322–5338.
78. S. Altun, O.N. Doruk, Şara, Effect of Al_2O_3 Nanoparticle Dispersion on the Thermophysical Properties of [EMIM][EtSO4] Ionic Liquid, International Journal of Thermophysics 44 (2023) 106.
79. B.C. Pak, Y.I. Cho, Hydrodynamic and Heat Transfer Study of Dispersed Fluids With Submicron Metallic Oxide Particles, Experimental Heat Transfer, 11 (1998) 151–170.
80. K.L. Main, B.K. Eberl, D. McDaniel, A. Tikadar, T.C. Paul, J.A. Khan, Nanoparticles Size Effect on Thermophysical Properties of Ionic Liquids Based Nanofluids, Journal of Molecular Liquids 343 (2021) 117609.
81. K. Oster, C. Hardacre, J. Jacquemin, A.P.C. Ribeiro, A. Elsinawi, Ionic Liquid-Based Nanofluids (Ionanocolloids) for Thermal Applications: an Experimental Thermophysical Characterization, Pure and Applied Chemistry 91 (2019) 1309–1340.
82. B. Bakthavatchalam, K. Habib, R. Saidur, N. Aslfattahi, S.M. Yahya, A. Rashedi, T. Khanam, Optimization of Thermophysical and Rheological Properties of Mxene Ionanocolloids for Hybrid Solar Photovoltaic/Thermal Systems, Nanomaterials. 11 (2021) 11020320.
83. U. Domańska, A. Rękawek, A. Marciniak, Solubility of 1-Alkyl-3-Ethylimidazolium-Based Ionic Liquids in Water and 1-Octanol, Journal of Chemical & Engineering Data 53 (2008) 1126–1132.
84. H.L. Ngo, K. LeCompte, L. Hargens, A.B. McEwen, Thermal Properties of Imidazolium Ionic Liquids, Thermochimica Acta 357–358 (2000) 97–102.
85. C. Villagrán, M. Deetlefs, W.R. Pitner, C. Hardacre, Quantification of Halide in Ionic Liquids Using Ion Chromatography, Analytical Chemistry 76 (2004) 2118–2123.
86. C. Villagrán, C.E. Banks, C. Hardacre, R.G. Compton, Electroanalytical Determination of Trace Chloride in Room-Temperature Ionic Liquids, Analytical Chemistry 76 (2004) 1998–2003.
87. D. Bejan, N. Ignatev, H. Willner, New Ionic Liquids With the Bis[Bis(Pentafluoroethyl) Phosphinyl]Imide Anion, [(C2F5)2P(O)]2N−Synthesis and Characterization, Journal of Fluorine Chemistry 131 (2009) 325–332.
88. K.R. Seddon, A. Stark, M.-J. Torres, Influence of Chloride, Water, and Organic Solvents on the Physical Properties of Ionic Liquids, Pure and Applied Chemistry 72 (2000) 2275–2287.
89. E. Gómez, N. Calvar, Á. Domínguez, Thermal Behaviour of Pure Ionic Liquids, In: Scott Handy (ed.), Ionic Liquids - Current State of the Art, InTech eBooks, 2015.

90. D.H. Zaitsau, M. Papusha, W.A. Kopp, K. Leonhard, Thermal Decomposition of 1-Ethyl-3-Methylimidazolium Acetate, Journal of Molecular Liquids 390 (2023) 123093.
91. J.J. Parajó, T. Teijeira, J. Fernández, J. Salgado, M. Villanueva, Thermal Stability of Some Imidazolium [NTf_2] Ionic Liquids: Isothermal and Dynamic Kinetic Study Through Thermogravimetric Procedures, The Journal of Chemical Thermodynamics 112 (2017) 105–113.
92. C. Niu, Y. Zhang, S. Ma, Y. Wan, H. Yang, X. Liu, An Intermediate Temperature Sodium Copper Chloride Battery Using Ionic Liquid Electrolyte and Its Degradation Mechanism, Ionics 25 (2019) 4189–4196.
93. H. Tokuda, K. Hayamizu, K. Ishii, Md. A.B.H. Susan, M. Watanabe, Physicochemical Properties and Structures of Room Temperature Ionic Liquids. 2. Variation of Alkyl Chain Length in Imidazolium Cation, The Journal of Physical Chemistry B 109 (2005) 6103–6110.
94. H.-B. Wu, B. Zhang, S.-H. Liu, C.-C. Chen, Flammability Estimation of 1-Hexyl-3-Methylimidazolium Bis(Trifluoromethylsulfonyl)Imide, Journal of Loss Prevention in the Process Industries 66 (2020) 104196.
95. J.M. Crosthwaite, M.J. Muldoon, J.K. Dixon, J.L. Anderson, J.F. Brennecke, Phase Transition and Decomposition Temperatures, Heat Capacities and Viscosities of Pyridinium Ionic Liquids, The Journal of Chemical Thermodynamics 37 (2005) 559–568.
96. D. Ji, Y. S. Kang, S.W. Kang, Accelerated CO_2 Transport on Surface of AgO Nanoparticles in Ionic Liquid BMIMBF4, Scientific Reports 5 (2015).
97. M. Knorr, M. Icker, A. Efimova, P. Schmidt, Reactivity of Ionic Liquids: Studies on Thermal Decomposition Behavior of 1-Butyl-3-Methylimidazolium Tetrafluoroborate, Thermochimica Acta 694 (2020) 178786.
98. V. Nozari, M. Zeeshan, S. Keskin, A. Uzun, Effect of Methylation of Ionic Liquids on the Gas Separation Performance of Ionic liquid/metal–organic Framework Composites, CrystEngComm 20 (44) (2018) 7137–7143.
99. H.-J. Liaw, C.-C. Chen, Y.-C. Chen, J.-R. Chen, S.-K. Huang, S.-N. Liu, Relationship between Flash Point of Ionic Liquids and Their Thermal Decomposition, Green Chemistry 14 (7) (2012) 2001.
100. F.S. Oliveira, A.B. Pereiro, J.M.M. Araújo, C.E.S. Bernardes, J.N.C. Lopes, S. Todorovic, G. Feio, P.L. Almeida, L.P.N. Rebelo, I.M. Marrucho, High Ionicity Ionic Liquids (HIIonic Liquids): Comparing the Effect of Ethylsulfonate and Ethylsulfate Anions, Physical Chemistry Chemical Physics 15 (41) (2013) 18138.
101. M.L. Williams, J.S. Dickmann, M.E. McCorkill, J.C. Hassler, E. Kiran, The Kinetics of Thermal Decomposition of 1-Alkyl-3-Methylimidazolium Chloride Ionic Liquids Under Isothermal and Non-Isothermal Conditions, Thermochimica Acta 685 (2020) 178509.
102. W. Wang, X. Ma, S. Grimes, H. Cai, M. Zhang, Study on the Absorbability, Regeneration Characteristics and Thermal Stability of Ionic Liquids for VOCs Removal, Chemical Engineering Journal 328 (2017) 353–359.
103. A. Jordan, N. Gathergood, Biodegradation of Ionic Liquids – A Critical Review, Chemical Society Reviews 44 (2015) 8200–8237.
104. E. Siedlecka, M. Czerwicka, J. Neumann, P. Stepnowski, J. Fernández, J. Thöming, Ionic Liquids: Methods of Degradation and Recovery, InTech eBooks, 2011.
105. Y. Cao, T. Mu, Comprehensive Investigation on the Thermal Stability of 66 Ionic Liquids by Thermogravimetric Analysis, Industrial & Engineering Chemistry Research 53 (2014) 8651–8664.
106. N. Koga, S. Vyazovkin, A.K. Burnham, L. Favergeon, N.V. Muravyev, L.A. Pérez-Maqueda, C. Saggese, P.E. Sánchez-Jiménez, ICTAC Kinetics Committee Recommendations for Analysis of Thermal Decomposition Kinetics, Thermochimica Acta 719 (2022) 179384.

107. A. Seeberger, A.-K. Andresen, A. Jess, Prediction of Long-Term Stability of Ionic Liquids at Elevated Temperatures by Means of non-Isothermal Thermogravimetrical Analysis, Physical Chemistry Chemical Physics 11 (2009) 9375.
108. D. Bejan, H. Willner, N. Ignatiev, C.W. Lehmann, Synthesis and Characterization of Bis[Bis(Pentafluoroethyl)Phosphinyl]Imides, $M^{+}N[(C2F5)2P(O)]2^{-}$, M = H, Na, K, Cs, Ag, Me4N, Inorganic Chemistry 47 (2008) 9085–9089.
109. G.B. Appetecchi, M. Montanino, M. Carewska, M. Moreno, F. Alessandrini, S. Passerini, Chemical–Physical Properties of Bis(Perfluoroalkylsulfonyl)Imide-Based Ionic Liquids, Electrochimica Acta 56 (2010), 1300–1307.
110. S. Kakinuma, H. Shirota, Femtosecond Raman-Induced Kerr Effect Study of Temperature-Dependent Intermolecular Dynamics in Pyrrolidinium-Based Ionic Liquids: Effects of Anion Species, The Journal of Physical Chemistry B 123 (2019) 1307–1323.
111. A.R. Choudhury, N. Winterton, A. Steiner, A.I. Cooper, K. A. Johnson, In Situ Crystallization of Low-Melting Ionic Liquids, Journal of the American Chemical Society, 127 (2005) 16792–16793.
112. C.S. Stefan, D. Lemordant, P. Biensan, C. Siret, B. Claude-Montigny, Thermal Stability and Crystallization of N-Alkyl-N-Alkyl′-Pyrrolidinium Imides, Journal of Thermal Analysis and Calorimetry 102 (2010) 685–693.
113. A.J. Ward, A.F. Masters, T. Maschmeyer, Facile, High-Yielding Preparation of Pyrrolidinium, Piperidinium, Morpholinium and 2,3-Dihydro-1H-Isoindolinium Salts and Ionic Liquids from Secondary Amines, RSC Advances 4 (2014) 23327–23337.
114. D.R. MacFarlane, P. Meakin, J. Sun, N. Amini, M. Forsyth, Pyrrolidinium Imides: A New Family of Molten Salts and Conductive Plastic Crystal Phases, The Journal of Physical Chemistry B 103 (20) (1999) 4164–4170.
115. E. Hosseini-Bab-Anari, A.M. Navarro-Suárez, K. Moth-Poulsen, P. Johansson, Ionic Liquid Based Battery Electrolytes Using Lithium and Sodium Pseudo-Delocalized Pyridinium Anion Salts, Physical Chemistry Chemical Physics 21 (33) (2019) 18393–18399.
116. Y.K.J. Bejaoui, F. Philippi, H.-G. Stammler, K. Radacki, L. Zapf, N. Schopper, K. Goloviznina, K.A.M. Maibom, R. Graf, J.A.P. Sprenger, R. Bertermann, H. Braunschweig, T. Welton, N.V. Ignatev, M. Finze, Insights into Structure–Property Relationships in Ionic Liquids Using Cyclic Perfluoroalkylsulfonylimides, Chemical Science 14 (8) (2023) 2200–2214.
117. S. Sowmiah, V. Srinivasadesikan, M.-C. Tseng, Y.-H. Chu, On the Chemical Stabilities of Ionic Liquids, Molecules 14 (9) (2009) 3780–3813.
118. B. Jóźwiak, G. Dzido, A. Kolanowska, M. Dzida, S. Boncel, High Operational Functionality Nanofluids from Choline Chloride, Ethylene Glycol, and Multi-Walled Carbon Nanotubes – Economical Substitutes for Ionanocolloids, Journal of Molecular Liquids 409 (2024) 125551.
119. V.S. Patil, A. Cera-Manjarres, D. Salavera, C.V. Rode, K.R. Patil, A. Coronas, Influence of Silver Nanoparticles Morphologies on Density, Viscosity and Thermal Conductivity of Silver Nanofluids and Silver Ionanocolloids, Journal of Nanofluids 7 (2017) 246–257.
120. R. Nanda, A. Kumar, Phase Behavior, Diffusion, Structural Characteristics, and pH of Aqueous Hydrophobic Ionic Liquid Confined media: Insights into Microviscosity and Microporosity in the [C4C4im][NTf2] + Water System, The Journal of Physical Chemistry B 119 (2015) 1641–1653.
121. B. Bakthavatchalam, K. Habib, R. Saidur, B.B. Saha, K. Irshad, Comprehensive Study on Nanofluid and Ionanocolloid for Heat Transfer Enhancement: A Review on Current and Future Perspective, Journal of Molecular Liquids 305 (2020) 112787.
122. S.S. Raut, R. Singh, U.M. Lekhak, Naturally Occurring Nanoparticles (NONPs): A Review, Next Sustainability 3 (2024) 100037.

123. H.T. Phan, A.J. Haes, What Does Nanoparticle Stability Mean?, The Journal of Physical Chemistry C 123 (27) (2019) 16495–16507.
124. M. Chereches, D. Bejan, E.I. Chereches, A.A. Minea, Experimental and Analytical Investigation of the Convective Heat Transfer Potential of PEG 400 Based Nanocolloids With Al_2O_3 and ZnO Nanoparticles, Heat and Mass Transfer 59 (5) (2022) 875–890.
125. A.K. Worku, D.W. Ayele, N.G. Habtu, G.A. Melas, T.A. Yemata, N.Y. Mekonnen, M.A. Teshager, Structural and Thermal Properties of Pure and Chromium Doped Zinc Oxide Nanoparticles, SN Applied Sciences 3 (2021) 7.
126. M. Chereches, D. Bejan, E.I. Chereches, A.A. Minea, Experimental Studies on Several Properties of PEG 400 and MWCNT Nano-Enhanced PEG 400 Fluids, Journal of Molecular Liquids 356 (2022) 119049.
127. J. Alizadeh, M.K. Moraveji, An Experimental Evaluation on Thermophysical Properties of Functionalized Graphene Nanoplatelets Ionanocolloids, International Communications in Heat and Mass Transfer 98 (2018) 31–40.
128. S. Mukherjee, S. Paria, Preparation and Stability of Nanofluids—A Review, IOSR Journal of Mechanical and Civil Engineering 9 (2013) 63–69.
129. X. Li, C. Zou, A. Qi, Experimental Study on the Thermo-Physical Properties of Car Engine Coolant (Water/Ethylene Glycol Mixture Type) Based SiC Nanofluids, International Communications in Heat and Mass Transfer 77 (2016) 159–164.
130. L. Das, N. Aslfattahi, K. Habib, R. Saidur, K. Irshad, S.M. Yahya, K. Kadirgama, Improved Thermophysical Characteristics of a New Class of Ionic Liquid + Diethylene Glycol/Al_2O_3 + CuO Based Ionanocolloid as a Coolant media for Hybrid PV/T System, Thermal Science and Engineering Progress 36 (2022) 101518.
131. G. Yıldız, Ü. Ağbulut, A.E. Gürel, A Review of Stability, Thermophysical Properties and Impact of Using Nanofluids on the Performance of Refrigeration Systems, International Journal of Refrigeration 129 (2021) 342–364.
132. E. Redel, R. Thomann, C. Janiak, First Correlation of Nanoparticle Size-Dependent Formation with the Ionic Liquid Anion Molecular Volume, Inorganic Chemistry 47 (1) (2007) 14–16.
133. C.M. Corrêa, M.A. Bizeto, F.F. Camilo, Direct Synthesis of Silver Nanoparticles in Ionic Liquid, Journal of Nanoparticle Research 18 (5) (2016).
134. P. Cao, Y. Li, Y. Wu, H. Chen, J. Zhang, L. Cheng, T. Niu, Role of Base Fluid on Enhancement Absorption Properties of Fe_3O_4/ionic Liquid Nanofluids for Direct Absorption Solar Collector, Solar Energy 194 (2019) 923–931.
135. M. Babucci, A. Akçay, V. Balci, A. Uzun, Thermal Stability Limits of Imidazolium Ionic Liquids Immobilized on Metal-Oxides, Langmuir 31 (33) (2015) 9163–9176.
136. K. Damodaran, Recent Advances in NMR Spectroscopy of Ionic Liquids, Progress in Nuclear Magnetic Resonance Spectroscopy 129 (2022) 1–27.
137. A.G.M. Ferreira, P.N. Simões, A.F. Ferreira, M.A. Fonseca, M.S.A. Oliveira, A.S.M. Trino, Transport and Thermal Properties of Quaternary Phosphonium Ionic Liquids and Ionanocolloids, The Journal of Chemical Thermodynamics 64 (2013) 80–92.
138. C. Hermida-Merino, A.B. Pereiro, J.M.M. Araújo, C. Gracia-Fernández, J.P. Vallejo, L. Lugo, M.M. Piñeiro, Graphene Ionanocolloids, Thermal and Structural Characterization, Nanomaterials 9 (11) (2019) 1549.

5 Heat Transfer Mechanisms in Ionanofluids

Marzena Dzida

5.1 THERMAL CONDUCTIVITY OF NANOPARTICLES AND LIQUIDS

Since nanofluids are composed of solid nanoparticles dispersed in a consistent liquid medium, understanding their thermal conductivity requires an in-depth analysis of the heat transfer behavior of both phases individually. Thermal conductivity in nanoparticles is governed by the way energy carriers, such as phonons or electrons, transport heat through or across the particles.[1] In non-metallic nanoparticles, heat is primarily carried by phonons, which are quantized vibrations of the crystal lattice structure. At the nanoscale, the high surface-to-volume ratio leads to frequent phonon scattering at surfaces, interfaces, and defects within the nanoparticles. Such scattering reduces the mean free path of the phonon and typically results in a lower thermal conductivity compared with the bulk material. However, in well-crystallized nanoparticles with minimal internal defects, phonon transport can remain relatively efficient. In metallic nanoparticles, heat is primarily conducted by free electrons. These electrons can efficiently transfer thermal energy within the nanoparticle core, but their motion is obstructed by scattering at the particle surfaces, grain boundaries, and interfaces with other materials. This scattering reduces the electron mean free path and causes the thermal conductivity to deviate from the bulk behavior. The thermal conductivity can be increased due to electron–phonon interactions between the nanoparticles and the base liquid.[2] Phonon energy is the energy that passes through atoms as vibrational energy. The size and shape of the nanoparticles also influence thermal conductivity. Smaller particles have a larger surface area, which increases interactions with the surrounding environment but also leads to greater phonon scattering. However, anisotropic shapes such as nanorods or nanotubes can promote directional heat transport and create efficient conductive pathways. The thermal conductivity also depends on the structure of the solid. For example, amorphous carbon has a thermal conductivity of 1.6 W m^{-1} K^{-1}, but diamond and multi-walled carbon nanotubes (MWCNTs, d = 14 nm, l = 2.5 μm) have thermal conductivities equal to 900[1] and 3000 W m^{-1} K^{-1}[3] at 298.15 K, respectively. Furthermore, thermal conductivities of single-layered graphene,[4] single-walled carbon nanotubes (SWCNTs, d = 1.7 nm, l = 2.6 μm),[3] hexagonal boron nitride (70–80 nm),[5] graphite (400–1200 nm),[5] Al2O3 (d <50 nm),[6] mesoporous carbon (average pore size 20–40 nm)[7] nanoparticles at 298.15 K that are loaded in ionanofluids are

DOI: 10.1201/9781003595250-5

equal to 1600–4000, 3500, 874.1, 35.7, 36, and ~12 W m^{-1} K^{-1}, respectively, whereas thermal conductivities of liquids range from ~0.06 W m^{-1} K^{-1} for fluorocarbons to ~0.6 W m^{-1} K^{-1} for water.[1] For ionic liquids, thermal conductivities range from 0.118 to 0.202 W m^{-1} K^{-1} for 1-butyl-1-methylpyrrolidinium bis(trifluoromethylsulfonyl) imide ([C_4C_1pyr][NTf_2]) and 1-ethyl-3-methylimidazolium dicyanamide ([C_2C_1im] [$N(CN)_2$]), respectively, at 298.15 K.[8]

5.2 IMPACT OF TYPE OF NANOPARTICLES ON THERMAL CONDUCTIVITY OF IONANOFLUIDS

The addition of nanoparticles enhances thermal conductivity with respect to ionic liquids, and the thermal conductivity is an increasing function of the nanoparticles' concentration. Exemplarily, the impact of the type of nanoparticle on thermal conductivity of [C_4C_1pyr][NTf_2]-based ionanofluids and the effect of morphology of carbon nanomaterials on thermal conductivity of 1-ethyl-3-methylimidazolium thiocyanate [C_2C_1im][SCN]-based ionanofluids at 298.15 K are presented in Figure 5.1.

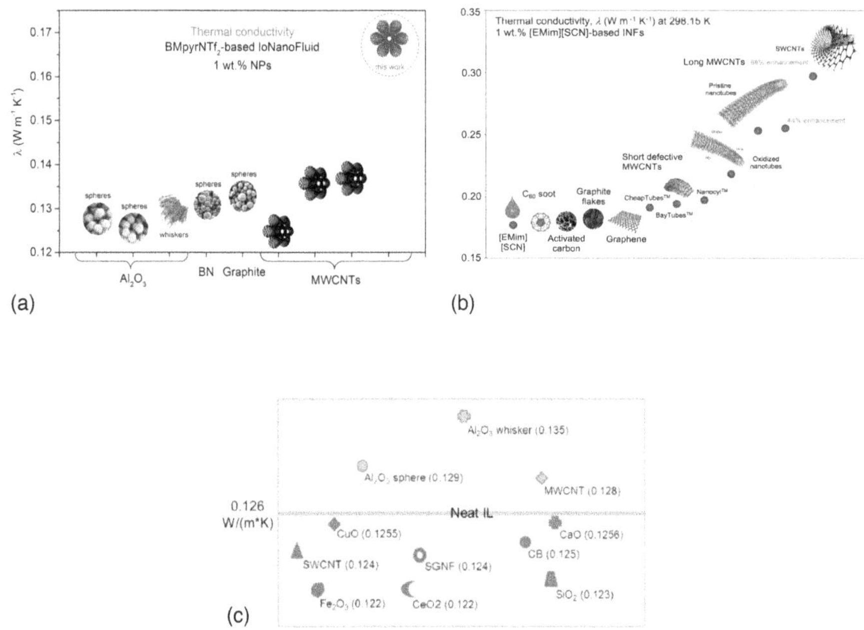

FIGURE 5.1 Impact of nanoparticles loading on thermal conductivity, λ, of [C_4C_1pyr] [NTf_2]-based ionanofluids[6,7,10,11] (this work[12]) (a); effect of morphology of carbon nanomaterials on thermal conductivity of [C_2C_1im][SCN]-based ionanofluids at 298.15 K[13] (b); thermal conductivity of ionanofluids containing [$C_4C_1C_1$im][NTf_2] + 0.5 wt.% nanoparticles at 323.15 K (c). (Copyright © 2022 The Authors. Published by the American Chemical Society. Publication[12] is an open-access under Creative Commons CC-BY 4.0 license. (Reprinted with permission from Fox, E. B.; Visser, A. E.; Bridges, N. J., Amoroso, J. W., Thermophysical Properties of Nanoparticle-Enhanced Ionic Liquids (NEILs) Heat-Transfer Fluids, *Energy Fuels*, **2013**, 27, 3385–3393. Copyright 2013 American Chemical Society.)

The SWCNTs and MWCNTs most significantly affected the thermal conductivity of ionanofluids. In turn, among several ceramic oxides, metal, and carbon-based materials dispersed in 1-butyl-2,3-dimethylimidazolium bis(trifluoromethylsulfonyl) imide ($[C_4C_1C_1im][NTf_2]$), only alumina and MWCNTs produced an increased thermal conductivity (Figure 5.1c).[9]

Fox et al.[9] examined ionanofluids with $[C_4C_1C_1im][NTf_2]$ and several nanoparticles (see Figure 5.1c). They noticed that the shape of the nanoparticle has a large influence on the temperature dependence of thermal conductivity. The increase in thermal conductivity of ionanofluid with whisker-shaped Al_2O_3 is much higher than that for spherical-shaped Al_2O_3 ones at temperatures 298.15–333.15 K. While the MWCNT-based ionanofluid initially has a thermal conductivity 4% higher than ionic liquid, but with increasing temperature, thermal conductivity rapidly decreases to that of ionic liquid, so that no benefit from the addition of the nanoparticle is observed. The SWCNT-based ionanofluid also characterizes a slight decline in thermal conductivity over the studied temperature range. The second type of ionanofluids is those obtained by the one-step method. Wang et al.[14] found that thermal conductivity of ionanofluid, obtained by one-step method, composed of $[C_4C_1im][PF_6]$ and $1.02{\cdot}10^{-3}$ vol.% Au (5.2 ± 1.2 nm), stabilized by cetyltrimethylammonium bromide, is 13.1% higher than that of $[C_4C_1im][PF_6]$ at 354.15 K but without visible changes at 306.15 K. This effect is probably connected to micro-convection induced by the Brownian motion of Au nanoparticles.

5.3 IMPACT OF STRUCTURE OF IONIC LIQUIDS ON THERMAL CONDUCTIVITY OF IONANOFLUIDS

In order to examine the influence of the structure of ionic liquids on the thermal conductivity of ionanofluids, Paredes et al.[15] compared the enhancement of thermal conductivity of ionanofluids with $[C_4C_1im]$-based ionic liquids and 1 wt.% MWCNTs Baytubes®C150HP (l = 1–10 μm, d = 13–16 nm) with thermal conductivity of ionic liquids (Figure 5.2). The thermal conductivity enhancement for ionanofluids decreases as presented in Figure 5.2a, while the thermal conductivity of the corresponding ionic liquid varies randomly (Figure 5.2b). For ionic liquids with

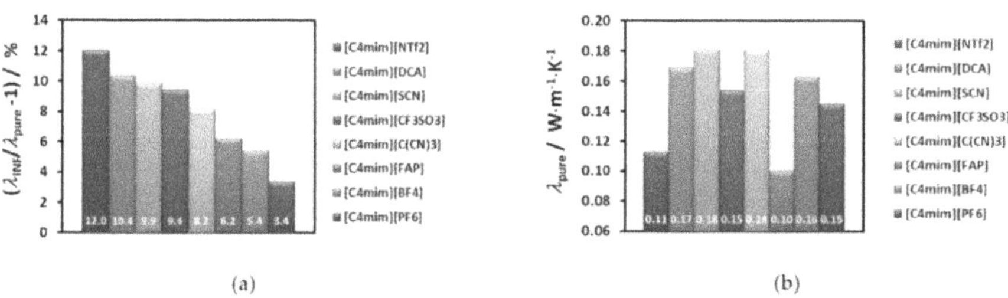

FIGURE 5.2 Thermal conductivity enhancement for ionanofluids composed of 1 wt.% MWCNTs Baytubes®C150HP and ionic liquids with the same cation $[C_4mim]$ in (a), and thermal conductivity of the appropriate ionic liquids in (b), at 298.15 K. (Publication[15] is an open-access under Creative Commons CC-BY license.)

$[C_4C_1im]^+$, the thermal conductivity increases as follows: $[N(CN)_2]^- > [C(CN)_3]^- \approx [SCN]^- > [BF_4]^- > [PF_6]^- > [CF_3SO_3]^- > [(CF_3SO_2)_2N]^- > [(C_2F_5)_3PF_3]^-$.[15] A bulky ion like $[(C_2F_5)_3PF_3]^-$, which is much less mobile, seems to contribute to a very low thermal conductivity. Cyano-anions introduce a much higher value for the thermal conductivity and a stronger dependence on temperature. The thermal conductivity of tetrafluoroborate ionic liquid $[C_4C_1im][BF_4]$ is high, though lower than that of cyano-ionic liquids, particularly at lower temperatures.[15] In general, thermal conductivity increases as the size of the anion decreases. Smaller anions typically exhibit better mobility, which enhances ionic interactions and, in turn, facilitates more efficient heat transfer pathways. Bulky anions such as $[NTf_2]^-$, which exhibit significantly lower mobility, tend to result in markedly reduced thermal conductivity. Figure 5.3 shows that cyano-anions introduce a much higher value to the thermal conductivity of ionanofluids.[15]

However, Figure 5.3b clearly shows that the enhancement of ionanofluid with $[C_4C_1im][NTf_2]$ is higher than that for ionanofluids with $[C_nC_1im]^+$ (n = 2, 6, 8) for similar weight loads of MWCNTs, despite the comparable thermal conductivity of $[C_nC_1im][NTf_2]$. This behavior was also observed by Ribeiro et al.[16] and Cwynar et al.[17] The same effect was found for ionanofluids with hexafluorophosphate ionic liquids $[C_4C_1im][PF_6]$ and $[C_6C_1im][PF_6]$.[10] Ribeiro et al.[16] explained this behavior by an increase in the entropy of the system from $[C_2C_1im]^+$ to $[C_4C_1im]^+$, due to the bigger contribution of the rotational/vibrational degrees of freedom, followed by a decrease in hydrogen bonding capacity of the bigger alkyl chain from $[C_4C_1im]^+$ to $[C_6C_1im]^+$. Nieto de Castro et al.[10] and Ribeiro et al.[16] suggested that a possible explanation was nanostructural organization in ionanofluids, the same as found for pure $[C_nC_1im][NTf_2]$ and $[C_nC_1im][PF_6]$ using molecular dynamics simulations.[18] Lopes and Pádua[18] found that ionic liquids containing cations with butyl or longer alkyl chains exhibit aggregation of these chains into nonpolar domains. This leads to the formation of a 3D network of ionic channels formed by anions and the imidazolium head of the cations, indicating microphase separation between polar and nonpolar domains within the liquid structure. The CNTs dispersed in the ionic liquid prefer to interact with the nonpolar domains formed by the alkyl chains, leading to the

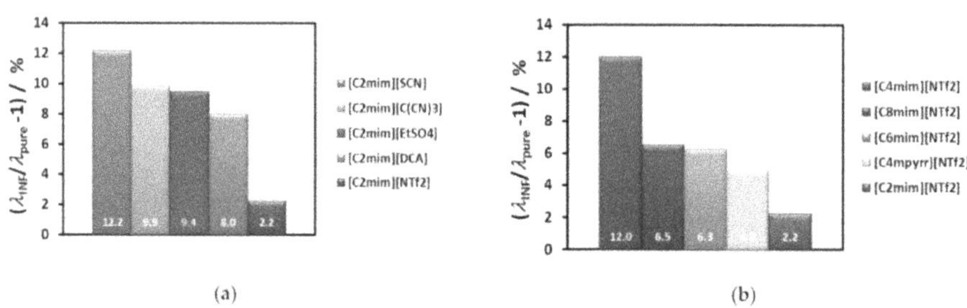

FIGURE 5.3 Thermal conductivity enhancement for ionanofluids composed of 1 wt.% MWCNTs Baytubes® C150HP and ionic liquids with the same cation [C²mim] in (a) and with the same anion $[NTf_2]$ in (b), at 298.15 K (Publication[15] is an open-access under Creative Commons CC-BY license.)

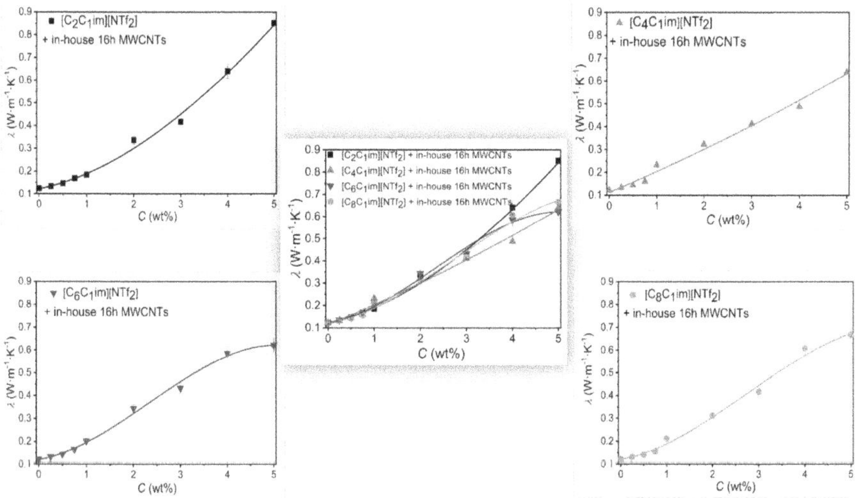

FIGURE 5.4 Thermal conductivity of ionanofluids at 298.15 K.[17,21,22] (Publications[21,22] are open-access under Creative Commons CC-BY 4.0 license.)

formation of microclusters that enhance heat transfer. Probably, for ionic liquids with the butyl chain, this microreorganization is optimal for CNTs. However, the structure of the ionanofluids is different compared with that of ionic liquids. The mechanism of thermal energy transfer in ionanofluids, reflected in macroscopic thermal conductivity, is minimally influenced by the properties of ionic liquids,[15] i.e., rather subtle effects can be observed, particularly in the nanoparticles concentration up to 1 wt.%. However, the thermal energy transfer may depend on the interactions between ions and the nanomaterial surface.[5,13,15,19] For comparison Figure 5.4 presents unrivalled thermal conductivity of ionanofluids based on in-house 16h synthesis,[20] originally 770-μm-long and multi-walled (several-dozen-walled, d_{outer} = 60–80 nm) CNTs (in-house 16h MWCNTs) studied by the Dzida and Boncel group.[17,21,22]

5.4 MECHANISMS OF THERMAL CONDUCTIVITY ENHANCEMENT IN IONANOFLUIDS

Generally, the incorporation of nanoparticles into ionic liquids resulted in unexpected enhancement of thermal conductivity. While the exact mechanism behind this enhancement remains under investigation, several hypotheses have been proposed to explain this phenomenon. Over the past few years, many publications tackled this issue through molecular dynamics calculations[23] and models starting from the classic Maxwell model[24] through its modifications and ending with the Murshed, Leong, and Yang (MLY) model.[25–27] Wang et al.[28] and Keblinski et al.[29] suggested possible mechanisms to explain the thermal conductivity enhancement in NFs, followed and extended for ionanofluids by Nieto de Castro et al.,[10] Murshed et al.,[27] Oster et al.,[7] Dzida et al.,[12,13] and França et al.[30] These include Brownian motion of nanoparticles, nanoconvection, interfacial nanolayer at the surface of nanoparticle, interphase particle/fluid structure, heat transport through the nanoparticles themselves, near-field radiation, nanoparticle clustering or aggregation,[16,29,31–39] thermal bridges formation,[13] and so-called zipping–unzipping–subzipping effects.[12] Additionally, the influence of types and size of nanoparticles,[7] nanoparticle surface chemistry, the nanoparticle morphology, and molecular perfection[12,13] on thermal conductivity of ionanofluids has been studied.

5.4.1 Theory of Thermal Conductivity of Interfacial Nanolayer

Recently, numerous theoretical investigations have been conducted to describe the thermal conductivity behavior of nanofluids, leading to the development of several models that account for different underlying mechanisms. These theoretical approaches can generally be classified into two main categories. The first category introduces the concept of a liquid–solid interfacial layer to model and explain the anomalous enhancement of thermal conductivity observed in nanofluids. The second category focuses on the dynamic contribution arising from the Brownian motion of nanoparticles in the fluid. In the case of ionanofluids, only the first concept has been developed and applied.

Interfacial ordering of ionic liquids on the external surface of carbon nanotubes. Among various mechanisms, the formation of structured interfacial nanolayers has emerged as one of the most compelling explanations for improved thermal transport resulting in enhanced thermal conductivity, particularly in ionanofluids.[5,10,12,13,15,16,27,30] The interfacial nanolayer consists of ordered base liquid atoms or ions, as in the case of ionic liquids, clustered at the boundary of the solid nanoparticle.[40] The ions near the surface of nanoparticles interact with, or adsorb onto the particle surface, forming a layered structure, thereby increasing the local atomic order at the interface,[28] exhibiting characteristics of an organized, solid-like system.[5,12,13,19,28,29,37,38,41–43] As a result, the interfacial nanolayer is considered a key factor in enhancing the thermal conductivity of ionanofluids. On one hand, it serves as a thermal bridge between the bulk liquid and the solid nanoparticle, enhancing the thermal conductivity of nanofluids and ionanofluids.[44,45] On the other hand, it helps to form structures that are also referred to as thermal bridges between nanoparticles themselves. In this case, sufficiently long CNTs form 3D networks of CNTs in the whole ionanofluid volume, leading to so-called zipping–unzipping–subzipping effects.[12] Pensado et al.[46] reported molecular dynamics results of the structure of the interfacial nanolayer of the $[C_2C_1im][SCN]$ at a flat graphene surface, including analyses of the positional and orientational ordering of the ions near the surface and charge density profiles. Both anions and cations are found in the first interfacial nanolayer near the graphene surface with the thickness of one ion, i.e., of ~0.4 to 0.5 nm. Interestingly, this periodicity aligns with the findings of Atkin and Endres group,[47–49] who reported solvation force profiles for various ionic liquids confined between atomic force microscope tips and surfaces such as mica, silica, graphite, and metals. Near the surface, an excess of positive charge is observed, followed by a region of negative charge and then another of positive charge. These charge density fluctuations confirm that both cations and anions are present close to the graphene surface. Pensado et al.[46] observed the second and the third adsorption layer, but these layers correspond to an isotropic liquid at distances around 2 nm from the graphene surface. The application of the ionic liquid-carbon potential derived from mixing rules of Lennard-Jones force fields[50–53] shows less structured solvation nanolayers with several atoms from both anions and cations appearing at the same distances from the carbon surface.[30] However, specifically parameterized models based on quantum calculations, used by Pensado et al.[46] and then by França et al.[30] led to more pronounced interfacial ordering. França et al.[30] developed an atomistic interaction model for ionic liquids and graphene or SWCNT ($l = 8.17$ nm, $r = 0.69$ nm for (10,10) and $l = 8.16$ nm, $r = 0.48$ nm, for (7,7)), that goes beyond standard combining rules. Due to the complex, non-additive interactions between ions and polarizable π-systems, the ionic liquids were divided into three fragments: cation head-groups, alkyl side chains, and anions. In turn, Zhou et al.[54] pointed out three main requirements of the force field needed to describe the interactions between the graphene surface and imidazolium-based ionic liquid: the conductive nature of the graphene sheets, the potentially strong π–π interactions between the imidazolium cation and the graphene, and the nanometer-sized heterogeneities observed in the bulk ionic liquid of imidazolium-based ionic liquids due to agglomeration of the nonpolar alkyl chains. França et al.[30] focused on $[C_4C_1im]$-based ILs with four anions: $[N(CN)_2]^-$, $[SCN]^-$, $[C(CN)_3]^-$, and $[NTf_2]^-$, and built interaction models by evaluating van der Waals energies for different fragment orientations on graphene and SWCNT surfaces. Simulations were conducted using both SWCNTs and stacked graphene sheets at 363 K. França et al.[30] applied quantum calculations with a density functional suitable for noncovalent interactions to parameterize an interaction model, including van der Waals terms and atomic partial charges on the materials. Importantly, França et al.[30] assumed the existence of the perfect sp^2-carbon nanomaterials. Once obtained, the next step of this computational approach involves simulating the nanomaterial–ionic liquid interface to elucidate the specific molecular interactions and their role in facilitating heat transfer. This approach enabled the investigation of both structural properties – such as the organization of the liquid within the interfacial region – and dynamic properties, including residence and diffusion times.[39] It was verified that anions and cations are present in the first interfacial layer, interacting directly with the material. In the initial stage, the proposed approach involved adjustment of a site–site interaction potential model between each atom of the ionic liquid and graphene, with the objective of reproducing the energies obtained from theoretical chemistry calculations. The choice of the type of potential function that best reproduces the energy results led to an n–m Lennard-Jones type potential, with exponents n and m different from those commonly used for simple organic molecules (12-6). The usability of the parameters fitted to interaction energy points between the ionic liquid and a graphene sheet was evaluated by calculating the interaction energy between the liquid and nanotubes, without further adjustments to the model. A more modest set of calculations was also carried out to produce a distribution of electrostatic charges in graphene sheets and nanotubes, both of "zigzag" and "arm-chair" structure. In this way, a model was developed that more closely represents the interactions of several ionic liquids with nanocarbons of different structures. The organization of the ions at the solid–liquid interface demonstrated a greater proximity of the anion to the carbon surface. They found that ionic liquids formed an interfacial nanolayer, particularly with cyano-based anions, leading to enhanced structural organization near the nanomaterial surfaces. In the nanolayer region, ions and carbon nanomaterials interact among themselves, above and below the π-system of the nanomaterials. Attraction is stronger for cations than for anions, whereas anions show stronger attraction for the hydrogenated edges due to their local positive charge. Thus, both anions and cations were found in the first interfacial nanolayer interacting with the carbon nanomaterials, just as found by Pensado et al.[46] Paredes et al.[15] visualized the results of molecular dynamics simulations realized by França et al.,[30] describing the structure of SWCNT/ionic liquid interface for SWCNT and 1-butyl-3-methylimidazolium tricyanomethanide $[C_4C_1im][C(CN)_3]$ at 363 K (Figure 5.5). The interfacial nanolayer of the outside wall of CNT consists of two distinct regions: an initial void region without ions of ~0.2 nm thickness (Figure 5.5),

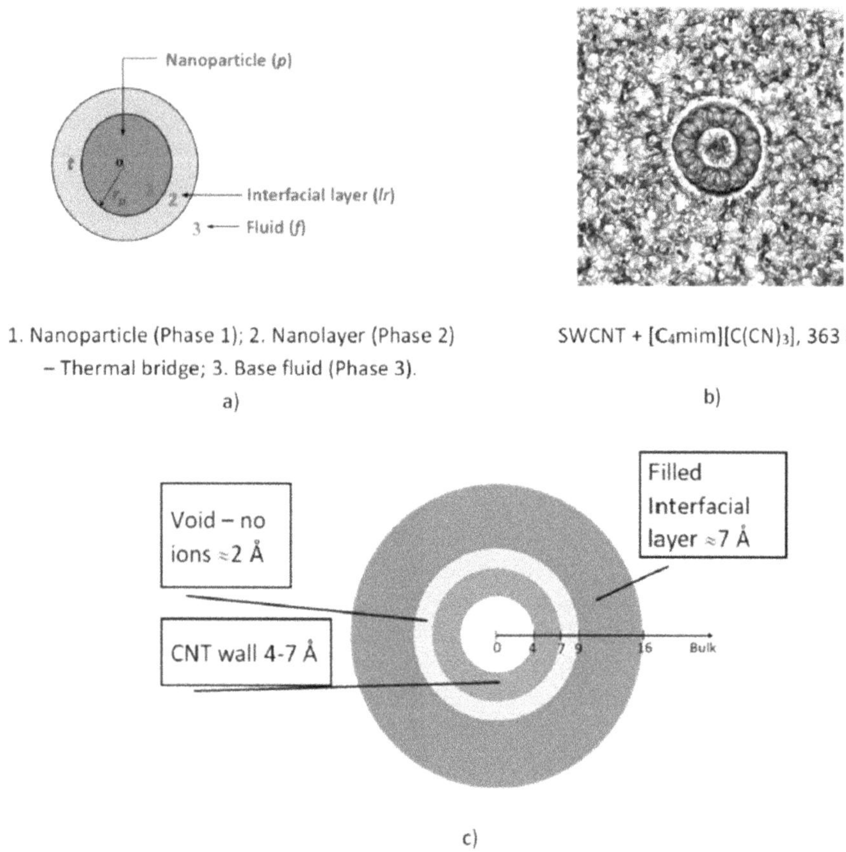

FIGURE 5.5 Schematic representation of interfacial nanolayer model for a spherical particle (a). Snapshot of simulation box with SWCNT+$[C_4C_1im][C(CN)_3]$ at 363 K. In this simulation, there are ions inside the CNT. The interfacial layers between the CNT walls and ionic liquid (in and out) are seen, also showing void spaces (no ions) (b). The simulation approximates the dimensions of SWCNT: $d_{inner} = 0.4$ nm, $d_{outer} = 0.7$ nm (c). Geometry is assumed to be similar along the axial dimension of the nanotube. (Publication is an open-access under Creative Commons CC-BY license.[15])

followed by a structured region where anions and cation head-groups are separated by ~0.037 nm. Naturally, in a real nanotube, the interfacial nanolayer forms an envelope along the tube's axial length, contributing to both axial and radial conduction effects. After the first nanolayer of ions surrounding the SWCNT, a depletion of anions is observed. In this region, the alkyl chains of imidazolium cations preferentially orient toward the bulk phase, reaching ~0.7 nm, and giving a total interfacial thickness of ~0.9 nm. This structural arrangement likely results in a local density of the ionic liquid lower than that of the bulk phase, meaning fewer molecules interact directly with the CNT surface. Consequently, the thermal conductivity across the interphase is reduced compared with the thermal conductivity of the bulk phase. These findings challenge many common assumptions about nanofluids based on molecular solvents.

At the interfacial nanolayers with graphene sheets, the ions are organized in two distinct layers of both cations and anions, with a third, outer, less organized layer where the alkyl chains are directed toward the bulk liquid, in the same way they were on the outside of the SWCNT. The density profiles reveal that the anions tend to align planar to the surface. The ordering of all studied cyano-based ionic liquids with respect to each carbon nanomaterial is similar. However, the flat and symmetric $[C(CN)_3]^-$ leads to stronger layering. The $[C_4C_1im][C(CN)_3]$ exhibits longer range charge ordering into the bulk phase, corresponding to more well-defined ionic layering. The structural results observed for $[C_4C_1im][NTf_2]$-based ionanofluids show similarities to those found in cyano-based ionanofluids, particularly in terms of the cations' side chains orientation toward the bulk phase. However, in interfacial layers adjacent to planar graphene surfaces, the side chains of the cations in $[C_4C_1im][NTf_2]$ are more likely to be located near the interface compared with cyano-based ionic liquids. This suggests that the structure of $[C_4C_1im][NTf_2]$ is softer than that of the cyano-based ionic liquids, forming solvation layers that are less rigidly defined by electrostatic charge ordering. This softer structure is attributed to its larger, more flexible anion, containing lower charge-density groups such as $-CF_3$. Franca et al.[30] further demonstrated that in $[C_4C_1im][NTf_2]$+SWCNTs systems, the interfacial region extends over a longer range than in cyano-based ionanofluids. Thus, in ionanofluids containing $[C_4C_1im][NTf_2]$, a thinner interfacial nanolayer is observed. This is in line with results obtained by the group led by Dzida and Boncel.[12,13,21,22] Dzida, Boncel, and coworkers, for the first time, experimentally found the interfacial nanolayer of ionic liquid on separate in-house 16h MWCNTs surface using transmission electron cryomicroscopy (cryo-TEM).[13] In the case of cyano-based ionanofluids, the individual nanotubes were covered with a few to several nanometers-thick ionic liquid layers. The average thickness of interfacial nanolayer on a separate nanotube is 4 ± 1.4[13]/8.4 ± 5.0 nm,[21] 10.1 ± 5.5 nm,[21] and 6.1 ± 3.8 nm[21] in ionanofluids with in-house 16h MWCNTs and $[C_2C_1im][SCN]$, $[C_2C_1im][N(CN)_2]$ or $[C_2C_1im][C(CN)_3]$, respectively (Figure 5.6). The interfacial nanolayer observed for ionanofluids with in-house 16h MWCNTs and $[C_3C_1pyr][NTf_2]$, or $[C_4C_1pyr][NTf_2]$, is substantially thicker and equals 25 ± 11 nm for both ionic liquids.[22] Dzida et al.[12] found that for $[C_4C_1pyr][NTf_2]$-based ionanofluids, the thickness of the interface nanolayer was practically independent of the nanotube type, i.e., equals 25 ± 11 nm for in-house 16h MWCNT and 27 ± 12 nm for Nanocyl™ NC7000 MWCNT ($l = 1.5$ μm, $d = 9.5$ nm)[12] (Figure 5.6).

These thicknesses of the interfacial nanolayer are significantly larger than those obtained by molecular dynamics simulations. However, they are of the same order

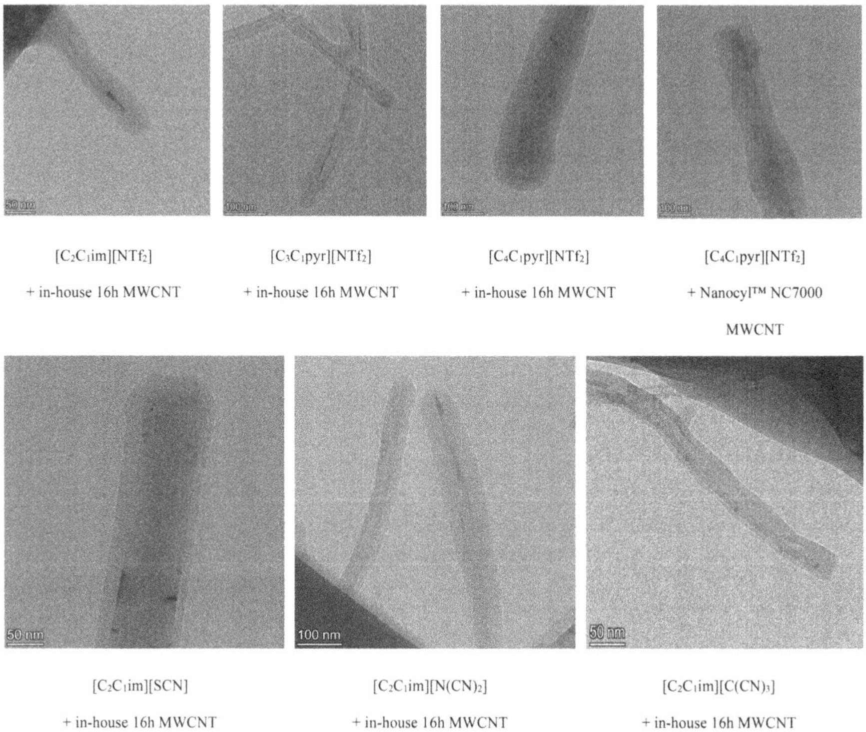

FIGURE 5.6 The cryo-TEM images revealing the thickness of the interfacial nanolayer in ionanofluids.[12,21,22] (Publications[12,21,22] are open-access under Creative Commons CC-BY 4.0 license.)

of magnitude as those obtained experimentally by Ribeiro et al.[16] Ribeiro et al.[16] were the first to obtain TEM images of an ionanofluid composed of [C$_4$C$_1$im][BF$_4$] and 1 wt.% MWCNTs Baytubes® C150HP. They found that the MWCNTs exhibit a random coil 3D structure, surrounded by a distinct white aureole, likely representing an interfacial nanolayer formed by [C$_4$C$_1$im][BF$_4$]. The average size of each structural unit ranges from 100 to 120 nm, with the interfacial nanolayer thickness estimated between 50 and 70 nm. Additionally, the coil aggregation was observed, which confirmed the formation of microclusters that should improve heat transfer paths. In turn, the Raman and IR analysis, as well as DFT and molecular dynamics calculations, show no strong interaction between imidazolium cations of [C$_4$C$_1$im][BF$_4$] or [C$_4$C$_1$im][PF$_6$] and SWCNTs.[55] Moreover, the overall organization of ionic liquids is not remarkably altered by SWCNTs. However, the shielding effect of ionic liquids on the π–π stacking interaction among SWCNTs takes the key role in dispersing the SWCNTs. Wang et al. found that the ionic liquids interact with SWCNTs through weak van der Waals interactions. This indicates that ionic liquids are ideal dispersing media for SWCNTs with no visible influence on the electronic structure and properties of SWCNTs[55].

Model of thermal conductivity in external interfacial nanolayer and bulk phase in ionanofluids with carbon-based nanomaterials. Based on the obtained structure of ionanofluids, França et al.[30] calculated the thermal conductivity of ionanofluid containing one SWCNT or one graphene stack in ionic liquid using non-equilibrium molecular dynamics. The thermal conductivity of the ionanofluids was calculated along the direction aligned with the longitudinal axis of the nanotube, including the areas where the nanotube is found, its edges, and also the surrounding ionic liquid, while these calculations were made along the perpendicular to the graphene planes. The resulting thermal conductivity contains a contribution coming from longitudinal heat transport, whereas those with the graphene sheet contain a contribution coming from perpendicular ones to the basal planes. The models used to calculate the interfacial thermal conductivity in suspensions with cylindrical or spherical nanoparticles are based on two assumptions. First, the interfacial thickness is ~ 1 nm for $[C_4C_1im][NTf_2]$ and between 0.84 and 0.89 nm for cyano-based ionic liquids. Second, the thermal conductivity of the interfacial nanolayer lies between that of the pure ionic liquid and the nanomaterial. The equilibrated simulation boxes and examples of temperature profiles for the ionic liquid + SWCNT and ionic liquid + graphene, respectively, are presented in Figure 5.7.[30] The enhancement of the thermal conductivity was calculated from two separate simulations: for the SWCNT + ionic liquid and for the pure ionic liquid. The heat flux was generated across the simulation box (z-axis) in both cases. In the ionic liquid + SWCNT systems, the heat flow

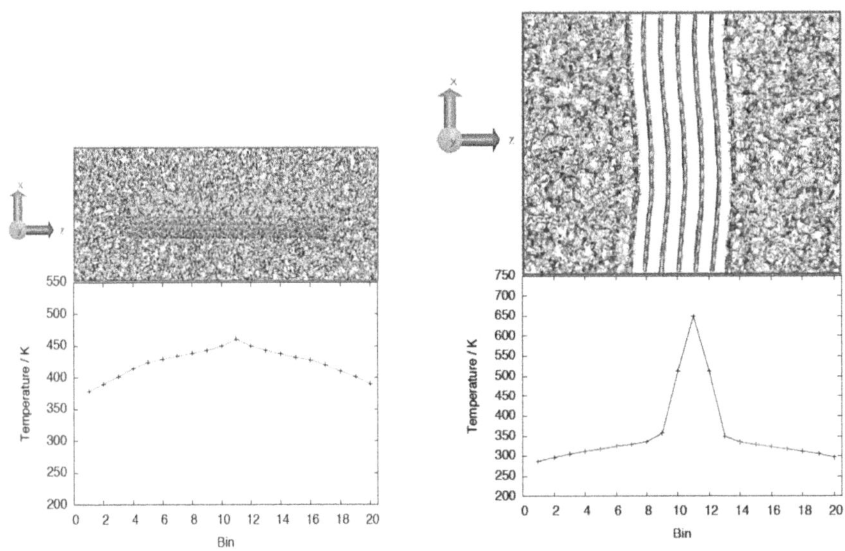

a) b)

FIGURE 5.7 (a) Snapshot of simulation box with $[C_4C_1im][SCN]$ + SWCNT (top) and temperature profile (bottom). Non-equilibrium molecular dynamics simulation at 423 K with exchange of Kinetic energy every 500 time steps; (b) Snapshot of equilibrated simulation box with $[C_4C_1im][N(CN)_2]$ + graphene (top) and temperature profile (bottom) with the interfacial zone in red. Non-equilibrium molecular dynamics simulation at 363 K with exchange of kinetic energy every 350 time steps. (Reproduced from Ref. [28] with permission from the PCCP Owner Societies.)

through the nanotube is favorable for heat transport and increases the overall thermal conductivity of the system (Figure 5.7a). There is a noticeable presence of two linear regimes in the temperature profiles. The local thermal conductivity for ionic liquid + graphene was calculated using the same values of the thickness of the interface as for SWCNT-based ionanofluid (Figure 5.7b). In this case, a perpendicular orientation to heat flow reduces the thermal conductivity.

Therefore, the thermal conductivity across graphene stacks was lower, but with local interface thermal conductivity enhancement of 15%–30%, depending on ionic liquid, due to directional dependence of heat flow.[30] Due to the significant anisotropy of thermal conductivity in graphene-based systems with respect to the direction of heat flux,[56] probably the local thermal conductivity at the interface also depends on the direction of the heat flow. Consequently, a perpendicular heat flow relative to the nanomaterial yields a lower enhancement. It is worth mentioning that the thermal conductivity anisotropy can reach up to 10^3 W m^{-1} K^{-1}, depending on whether heat is conducted in-plane or cross-plane, as observed in stacks of graphene sheets and graphite. Similar trends are also evident in perfect, defective, and composite systems.[30] Another mechanism to account for the difference between the thermal conductivity of ionanofluids is based on the Kapitza resistance.[57] This is a barrier to heat transfer associated with weak dispersion forces between the liquid and the carbon nanomaterials. França et al.[30] observed that temperature gradients remain continuous from the bulk ionic liquid to the interface in all examined ionanofluids, indicating the absence of exceptional interfacial discontinuity, i.e., no reasonable Kapitza length (Figure 5.8). Franca et al.[30] showed that the use of specifically parameterized potentials led to enhanced interfacial ordering and up to 24% increase in thermal conductivity for cyano-based ionic liquids + SWCNTs ionanofluids, as well as up to 34% for $[C_4C_1im][NTf_2]$+SWCNTs. Thus, the ionic liquid presenting a greater thermal conductivity enhancement in the ionic liquid + SWCNT systems is $[C_4C_1im][NTf_2]$, likely due to the longer range of the interfacial region, whereas the cyano-based ionanofluids characterize a smaller interfacial thickness and also a smaller thermal conductivity enhancement. However, the thermal conductivity of ionanofluids, calculated by Franca et al.[30] is higher than that experimentally reported. Franca et al.[30] stated that the presence of nanotube bundles in real ionanofluids explained the lower values of thermal conductivity. This effect was verified and studied in detail by Dzida et al.[12,13] whereas in the simulated systems, the SWCNT is isolated in the ionic liquid phases. The most important was the realization that the significant enhancement in thermal conductivity happens in the interfacial region, with no significant discontinuity in the temperature profile at the solid–liquid interface.[39]

Interfacial ordering of ionic liquids on the surface of metallic nanoparticles. In the case of ionanofluids with metal nanoparticles, the study of interfacial nanolayer formation is connected with the stability of these systems. In the literature, the electrostatic stabilization realized by the existence of an electric double layer was postulated.[58] In such an electric double layer, a first solvation shell of anions surrounds the metal cluster, followed by a less ordered layer of cations. It was pointed out that the Derjaugin–Landau–Verwey–Overbeek theory enabled the determination of the stability of colloids, considering a combination of Coulomb forces and van der Waals interactions.[58] However, the structure of the interfacial nanolayer has some similarities to non-metallic-based ionanofluids. The molecular dynamics simulations for solvation of a 2-nm ruthenium particle in $[C_4C_1im][NTf_2]$ reveal the electrostatic interface layer of 0.4–0.5 nm of both the cation head and anion.[58] The terminal atoms of the alkyl side chain of the imidazolium cations are found further away from the nanoparticle, at distances ~1.8 nm from its center. This is a result of the stronger attractive interaction of ions with metal surfaces. The $-CF_3$ groups of the anions are also directed away from the nanoparticles, which is in line with results for MWCNTs-based ionanofluids.[12] The imidazolium ring lies, most probably, perpendicular to the nanoparticle surface, with the $-CH_3$ group attached to the N3 closer to the metal. The interactions of the imidazolium ring with neighboring anions are thus maximized. These strong interactions have been observed by surface-enhanced Raman spectroscopy on gold nanoparticles.[59] Pensado and Pádua[58] concluded that in ionic liquids, metallic nanoparticles were preferentially solvated by the charged groups of the ions, forming an interfacial layer only one ion thick. As a result, both cations and anions are in direct contact with the nanoparticle surface, while nonpolar groups and side chains orient away from it. The stabilization mechanism of metallic nanoparticles in this medium is therefore not governed by a classical electrostatic double layer, nor is it attributable to steric hindrance from the alkyl chains of the cations, which in this case are too short to provide such an effect. The narrow size distribution of nanoparticles synthesized in situ, together with the observed correlation between particle size and characteristic structural length scales of the ionic liquid (such as the effective anion volume or the size of nanoscopic heterogeneities), demonstrates that these intrinsic length scales influence both the nucleation and growth of nanoparticles as well as the stability of the resulting colloids. Stabilization thus arises from the solvent acting as a structural matrix or template, rather than from electrostatic double layer or steric repulsion.

Interfacial ordering of ionic liquids inside carbon nanotubes. The molecular dynamics studies reveal that ionic liquids formed different structures not only on the external surface of CNTs. The different structures are also formed inside CNTs, which depend on the nanotube diameter.[30,46] Pensado et al.[46] examined $[C_2C_1im]$ [SCN]-based ionanofluids with CNTs with three different chirality indexes: (10,0), (15,0), and (20,0), with $d = 0.14–0.27$ nm and $l = 30$ nm. The three different systems were studied for each carbon nanotube: (*i*) the average mass density of the ionic liquid inside of the CNT is the same as the bulk density of the ionic liquid; (*ii*) the average mass density inside of the CNT is 0.5 times the bulk ionic liquid density; (*iii*) the average mass density is 1.2 times the density of the bulk phase. The presence of the CNT strongly modifies the structure of the ionic liquid. When an excess of ionic liquid enters the CNT, the density inside is larger than the density of the bulk liquid. For the (10,0) CNT, the ionic liquid is packed more compactly close to the surface, and a second region of high density appears in the center of the CNT. Therefore, even if the size of the ionic liquid studied here is small, there is a region where the local density is low. Increasing the size of the CNT makes it easy to accommodate the excess of ionic liquid. For the (15,0) CNT, the strong adsorption layer at the surface of the nanotube was observed, followed by a region of low density and an enlarged second solvation layer placed close to the center of the CNT, with the ionic liquid's structure being more disordered. A similar effect is observed for (20,0) CNT, the central region presents a lower level of ordering that resembles that of an isotropic liquid, but still, there are regions with low local density. When the number of ion pairs inside the CNT decreases, the free volume in the ionic liquid is redistributed. In the region with a strong adsorbed layer of ions near the CNT surface, there is a strong local density of the ionic liquid, whereas a void region appears in the center of the CNT. França et al.[30] found that in wider SWCNTs, such as (10,10), the alkyl chains of the cations are directed toward the center of the tube and form nonpolar domains at the core, while this arrangement is absent in narrower CNTs-like (7,7). The structural ordering within (10,10) and (7,7) CNTs is consistent for ionanofluids with cyano-based ionic liquids: in the narrower (7,7) tube, the cation head-groups are located near the axis, while in the wider (10,10) tube, the alkyl side chains occupy the central region. A notable feature in the (10,10) CNTs is the positioning of the cations' alkyl chain along the tube axis, with the local density of cyano-type groups decreasing in the following order $[SCN]^-$ > $[N(CN)_2]^-$ > $[C(CN)_3]^-$. This indicates that larger cyano-anions resulted in less ordered nonpolar domains at the core of the (10,10) CNT. Additionally, the $-CF_3$ groups in $[C_4C_1im][NTf_2]$, which represent the least charged regions of the anions, are found near the center of the (10,10) tube and adjacent to the inner wall of the narrower (7,7) tube. It is extraordinary that the nanosegregation postulated for pure ionic liquids[18] occurred inside a CNT. These effects remain to be elucidated in the context of the thermal conductivity of ionanofluids.

5.4.2 Zipping, Unzipping, and Subzipping Effects in Carbon Nanotube-Based Ionanofluids

The zipping, unzipping, and subzipping effects were discussed by Dzida and Boncel's group in the experimental and theoretical research.[12,21,22] Jóźwiak et al.[13] obtained the enhancement of the thermal conductivity of 67.8% for $[C_2C_1im][SCN]$ + 1 wt.% of Tuball SWCNTs (l = 5.0 μm, d = 2 nm) at 298.15 K. Dzida et al. found that the thermal conductivity of ionanofluids composed of $[C_3C_1pyr][NTf_2]$, $[C_4C_1pyr][NTf_2]$ and 1 wt.% of C-sp^2 rich, long, crystalline in-house 16h MWCNTs is higher than those of base ionic liquid by 39.4% and 44%, respectively, at 298.15 K.[22] Although thermal conductivity of ionanofluids composed of $[C_4C_1pyr][NTf_2]$ and 1 wt.% of sp^3-C rich MWCNTs, short Nanocyl™ NC7000 MWCNTs is higher than that of the base ionic liquid only by 14% at 298.15 K.[12] For the imidazolium-based ionanofluids series, a set of increases in thermal conductivity was observed at 1.0 wt.% in-house 16h MWCNTs. Starting from $[C_2C_1im][NTf_2]$-based INF with 48% increase, $[C_4C_1im][NTf_2]$-based ionanofluid showed 90% enhancement, with $[C_6C_1im][NTf_2]$-based ionanofluid having 66% increase, finishing with $[C_8C_1im][NTf_2]$-based INF with 73% increase in thermal conductivity at 298.15 K.[17] The thermal conductivity enhancement for ionanofluids with 1 wt.% of in-house 16h MWCNTs and $[C_2C_1im][SCN]$, $[C_2C_1im][N(CN)_2]$, or $[C_2C_1im][C(CN)_3]$ equals 36%, 19%, and 20% at 298.15 K, respectively.[21] Dzida et al.[12] carried out structural and spectroscopic studies, including optical microscopy, scanning electron microscopy (SEM), TEM, cryo-TEM, Raman spectroscopy, as well as molecular dynamics simulations correlated with experimental thermal conductivity of $[C_4C_1pyr][NTf_2]$-based ionanofluids. The studies focused on long in-house 16h MWCNTs with an originally high aspect ratio

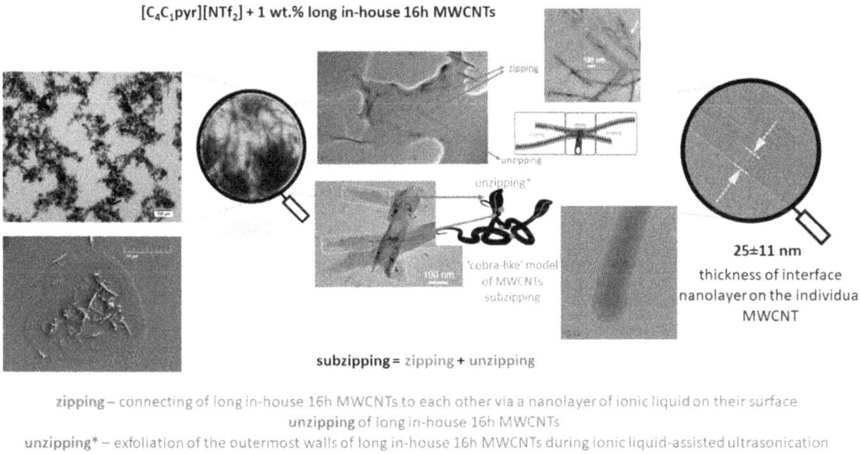

FIGURE 5.8 Subzipping effect for long MWCNTs-based ionanofluid.[12] (Publication[12] is an open-access under Creative Commons CC-BY 4.0 license.)

of 11,000, and as the reference system, commercially available, short, and defective Nanocyl™ NC7000 MWCNTs with a moderate aspect ratio of 150. The non-aromatic $[C_4C_1pyr]^+$ was selected to disable π–π interactions with MWCNTs and to allow accurate charge referencing. The $[NTf_2]^-$ was chosen because it is a relatively large, complex anion with a delocalized negative charge along the S–N–S ion core, and the steric hindrance reduces the number of ion–ion interactions. Dzida et al.[12] studied the way of interconnection through the nanolayers of the long in-house 16h MWCNTs (Figure 5.8) and short Nanocyl™ NC7000 MWCNT-based ionanofluids (Figure 5.9). The SEM analysis revealed that the long in-house 16h MWCNT-based ionanofluids formed highly fibrous and well-ordered droplets (Figure 5.8),

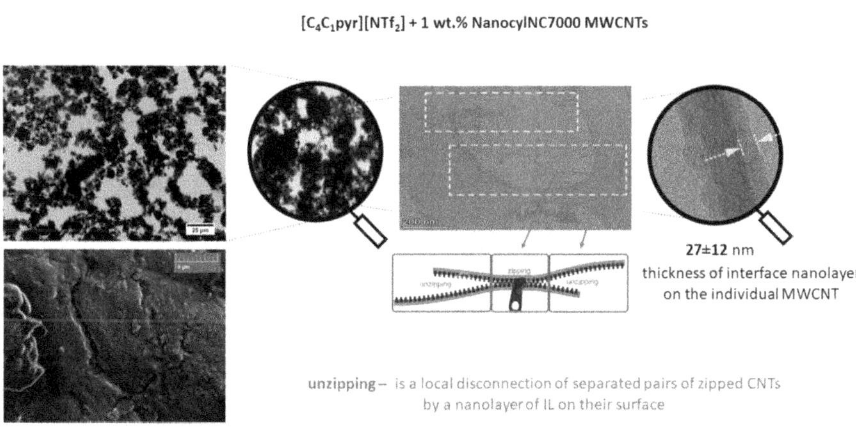

FIGURE 5.9 Zipping and unzipping effects for short MWCNTs-based ionanofluid.[12] (Publication[12] is an open-access under Creative Commons CC-BY 4.0.)

in contrast to the more isotropically distributed droplets observed with the shorter Nanocyl™ NC7000 MWCNTs (Figure 5.9). The light micrograph of long in-house 16h MWCNTs-based ionanofluids confirmed existing superstructures by long-range interactions. The in-house 16h MWCNTs formed compact networks of straight and bent nanotubes, constructing the interconnected superlattice vibration pathways. Therein, the physical intertube contact could generate steady thermal and mechanical bridges. The TEM and cryo-TEM imaging revealed the partially subzipping of long in-house 16h MWCNT networks in ionic liquid. This is a result of interactions between two neighboring nanotubes, which fragmentarily coalesce by the ionic liquid nanolayers, i.e., intertube zipping together with locally unzipped individual pairs of nanotubes and/or longitudinally unzipped nanotubes.

Zipping refers to the connection of CNTs to each other via a nanolayer of ionic liquid on their surface. Unzipping is a local disconnection of separated pairs of zipped CNTs by a nanolayer of ionic liquid on their surface or exfoliation of the outermost walls of CNTs during ionic liquid-assisted ultrasonication. A subzipping is various configurations of both zipped and unzipped CNTs (Figure 5.8).[12] The subzipped 3D nanotube networks of lengths from several hundreds of nanometers to a few microns (Figure 5.8) were found in ionanofluids with long MWCNTs. It is important that long in-house 16h MWCNTs are shortened due to sonication during sample preparation.[60] The TEM imaging revealed that the nanotube system represents a "cobra-like" model of subzipping, involving both individual exfoliated MWCNTs and MWCNT networks (Figure 5.8). In contrast, short Nanocyl™ NC7000 MWCNTs cannot form subzipped nanotube bundles into the 3D networks (Figure 5.9). Separated individual pairs of zipped and locally unzipped short nanotubes can be observed (Figure 5.9). In order to study the architecture of subzipped nanotube bundles in molecular dynamics simulations, three layers of ions were defined (Figure 5.8). Layers I and III contain ions that face the bulk phase, and layer II, where the ions are located between the two carbon sheets, mimicking the outermost CNT walls. The organization of ions in layers I and III was similar. When the number of ionic liquid pairs between carbon sheets was such that only one layer of ions was formed, the ions produced a cohesive and ordered structure (Figure 5.10). In turn, when the ionic liquid layer faced the bulk phase, a loose structure was formed, and the molecules could easily swap positions near carbon sheets (Figure 5.10). This dynamic also led to the momentaneous formation of "holes" in this layer (yellow areas in Figure 5.10). In the central layer, ions are aligned parallel to the surface, suggesting that they are restrained to in-plane movements, while in layers I and III, the molecules exhibit translational and rotational degrees of freedom (Figure 5.10). Therefore, the cohesive nature of layer II, located between the two carbon sheets, leads to a more efficient and stable charge balance of the ions close to the surface. Hence, considering that this process is essentially controlled by internal energy, ΔU, the obtained results $U_{\text{int}} = -38.2$ kJ mol^{-1} suggest that the most negative U_{int} characterized formation of a structure where adjacent CNTs are separated by one layer of ionic liquid ions, i.e., distance of carbon sheets, $d_{\text{CS}} = 0.76$ nm.

Thus, the simulation results indicate that the formation of such a structure is favorable from a thermodynamic point of view. Studying the character of MWCNT-ionic liquid interactions via Raman spectroscopy from the "CNT perspective" (Figure 5.10),

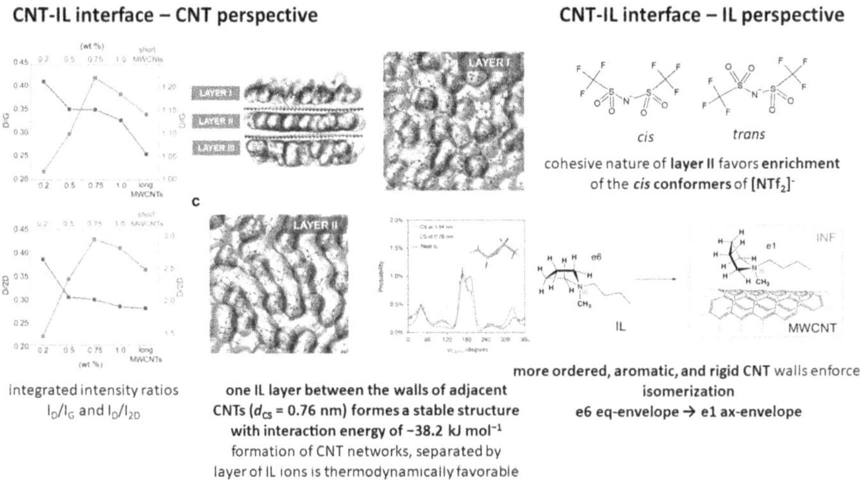

FIGURE 5.10 The organization of ionic layers for $[C_4C_1pyr][NTf_2]$ + MWCNTs.[12] (Publication[12] is an open-access under Creative Commons CC-BY 4.0 license.)

an upshift was encountered of the prominent Raman bands as a function of the MWCNT loading (see Figure 6a,b in Ref. [12]). This effect correlates with the bundle-penetrating power of ionic liquid, weakening tube–tube interactions via subzipping of the long in-house 16h MWCNT networks. Importantly, the in situ formation of CNT bundles has characteristic synthesis features. Long in-house 16h MWCNTs, cut during sonication, are quenched using covalent bonds involving cations. The "new" MWCNTs undergo subzipping and become functionalized at their tips and edges. Thus, interactions between the individual CNTs are partially changed from π–π stacking into more complex ones through functionalized "contacts", enabling subzipping of the CNT network. Analysis of the integrated intensity ratios I_D/I_G and I_D/I_{2D} revealed distinct modification patterns in MWCNTs (Figure 5.10). For long in-house 16h MWCNTs, both ratios decreased with concentration, indicating increased functionalization linked to CNT subzipping. In contrast, the more defective, short Nanocyl™ NC7000 MWCNTs showed increasing ratios at lower concentrations due to their higher C-sp³ content. In both types, covalent bonds were formed between CNT walls and ionic liquid cations, for long MWCNTs with "cation capture" potentially neutralizing ultrasonication-induced dangling bonds via in situ methylation and tertiary amine formation.[61] Notably, ionic liquid-induced modifications differed by MWCNT type: in short MWCNTs, outer wall exfoliation at low concentrations exposed more graphitized layers, while higher concentrations led to a defect peak due to radical neutralization. For long MWCNTs, exfoliation was minimal, and both cutting and functionalization levels decreased with increasing ionanofluid viscosity, reflecting their more aromatic, graphitized character. A slightly lower upshift of all Raman bands for the long MWCNTs resulted from subzipping. In total, long MWCNTs were less "debundled", whereas short, sp³-defectuous MWCNTs emerged as more strongly interacting in all aspects with the ionic liquids. Strong band reorganization of Raman spectra at approximately 3200–3000 cm^{-1} and a slight modification in the 3000–2800 cm^{-1} region suggest structural changes within the pyrrolidinium head and insignificant alteration of the butyl tail in ionanofluids (see Figure 7b,d in Ref. [12]), respectively. The presence of only two unique Raman modes at 934 and 909 cm^{-1} for the reference ionic liquid spectrum corresponds to the e6 eq-envelope cation conformers (see Figure 7a,c in Ref. [12]). Nevertheless, both MWCNT-based ionanofluids manifested the additional band located at a lower frequency that corresponds to the e1 ax-envelope cation isomer with the butyl chain in the axial position relative to the ring plane. The mutual intensity between the e1- and e6-bands was independent of the MWCNT content for short MWCNT-based ionanofluids, while it was rising for the long MWCNT-based ionanofluids. The more ordered, aromatic, and rigid CNT walls enforce the e6→e1 isomerization, which becomes particularly favorable for a considerable area of well-developed graphene walls in ionanofluids. The 800–700 and 480–260 cm^{-1} regions prove *trans* conformation of the $[NTf_2]^-$ as the dominating one (with a limited number of "free" *cis* conformation). This effect was found for the ionic liquid and short MWCNT-based ionanofluids regardless of the CNT content (see Figure 8a in Ref. [12]). The ionanofluids with long MWCNTs tended, in a more pronounced manner, to equalize the population of *trans* conformers and *cis* conformers of $[NTf_2]^-$ by hampering the free rotation in the cation conformation, caused by its *in situ* covalent functionalization with the CNT surface. This reflects the increase in the intensity of bands in the 340–260 cm^{-1} range compared with the 480–340 cm^{-1} region (see Figure 8c in Ref. [12]). The molecular dynamics confirmed that the cohesive nature of layer II, imposed by the close carbon sheets, favors the enrichment of the *cis* conformer of $[NTf_2]^-$ in ionanofluid, and the alkyl chain of $[C_4C_1pyr]^+$ revealed no conformational change with the distance between carbon sheets. The presented relationship between the internal structure and thermal conductivity of ionanofluids is in line with the results reported by Franca et al.[30] Moreover, this study specified the bundle effect postulated by Franca et al.[30]

5.4.3 Percolation Theory

The subzipping effect can be in some sense related to percolation one. The percolation theory suggests that the thermal conductivity enhancement is due to the formation of a "percolation" network of nanoparticles within the nanofluid.[2] When the concentration of nanoparticles is low, they are dispersed randomly in the base liquid and do not contribute to thermal conductivity. However, as the concentration of nanoparticles increases, they begin to aggregate to form a network that enhances thermal conductivity (see Figure 5.11). However, after exceeding the concentration characteristic for a given system, these aggregated clusters lead to reduced Brownian particle velocity within the nanofluid.[62]

5.5 Mathematical Models of Thermal Conductivity of Ionanofluids

To an interpretation of the thermal conductivity of ionanofluids, several mathematical models were applied. Here, a brief overview of only those models that have been used to calculate the thermal conductivity of ionanofluids is presented.

FIGURE 5.11 Scheme of the MWCNT (l = 1.5 μm, d = 10 nm) network evolution at increasing filler concentration in glycerine. First scheme for MWCNTs loading < 0.046 wt.% related to electrical percolation; the second scheme for MWCNTs loading 0.046 < φ_{wt} < 0.16 wt.% related to electrical and mechanical percolation; above the mechanical percolation (φ_{wt} > 0.16 wt.%), the system presents a solid-like behavior. (Publication[62] is an open-access under)

5.5.1 Maxwell Model

The effective medium theory is a theoretical framework that is often used to predict the properties of nanofluids. The basic idea behind the effective medium theory is that the properties of a composite material can be approximated by averaging the properties of the individual components. In nanofluids, the individual components are the base liquid and the nanoparticles suspended within it. The effective medium theory is typically used to estimate the thermal conductivity of nanofluids based on the properties of the base liquid and nanoparticles.[2] The earliest model was derived by Maxwell[24] from the definition of the effective thermal conductivity, λ_{eff}, of a two-component mixture and is expressed as follows:

$$\lambda_{eff} = \frac{\lambda_{NP}\varphi_{NP}\left(\frac{\partial T}{\partial x}\right)_{NP} + \lambda_{IL}\varphi_{IL}\left(\frac{\partial T}{\partial x}\right)_{IL}}{\varphi_{NP}\left(\frac{\partial T}{\partial x}\right)_{NP} + \varphi_{IL}\left(\frac{\partial T}{\partial x}\right)_{IL}}, \tag{5.1}$$

where $(\partial T/\partial x)$ is the overall average temperature gradient, φ is the volume fraction, and the subscripts NP and IL refer to nanoparticle and ionic liquid.

The application of the effective medium approximation for randomly dispersed and uniformly sized spherical particles leads to the following relation:

$$\frac{(\partial T/\partial x)_{NP}}{(\partial T/\partial x)_{IL}} = \frac{n\lambda_{NP}}{\lambda_{IL} + (n-1)\lambda_{NP}}, \tag{5.2}$$

where n is a shape factor of the dispersed particles [5]. Maxwell assumed the existence of spherical particles with n = 3. The thermal conductivity of an ionanofluid based on the Maxwell model is expressed as follows:

$$\lambda_{INF} = \lambda_{IL}\frac{\lambda_{NP} + 2\lambda_{IL} + 2\varphi_{NP}(\lambda_{NP} - \lambda_{IL})}{\lambda_{NP} + 2\lambda_{IL} - \varphi_{NP}(\lambda_{NP} - \lambda_{IL})}. \tag{5.3}$$

The mechanistic Maxwell model can be applied for systems with uniform dispersion of spherical particles with no particle interactions, i.e., for dilute suspensions of spherical particles,[63] neglects the dynamic state of the particles, and does not consider the interaction between the particles and the base liquid.[2] The Maxwell model[63] as well as Maxwell-type models[1] predicted thermal conductivity of nanofluids of 0%–15% higher than that of a base liquid when $\phi_{NP} \leq 0.05$.[63] Turian et al.[64] demonstrated that these models were able to fit experimental data for nanofluids within 2 vol.% when $0.4 < (\lambda_{NP}/\lambda_{liquid}) < 2.4$. Agreement with experiment becomes less satisfactory as $(\lambda_{NP}/\lambda_{liquid})$ increases and also as ϕ_{NP} increases. The increasing ϕ_{NP} leads to aggregation of the particles, and this effect cannot be accounted for by this model. It should be emphasized that it is not correct to add up quantities of substances in different phases. In this case, it is not correct to add up the thermal conductivity of a solid nanoparticle and the thermal conductivity of an ionic liquid, which is in the liquid state (see Eq. (5.1)). In the case of solutions with a solid solute, a non-symmetric reference system is used. In the case of nanofluids, it is a more complicated issue, because the final system is composed of solid and liquid phases. Therefore, the simple mixing rules are not suitable here. Furthermore, in this context, the nomenclature regarding nanofluids should be clarified in the literature. Nanofluid is composed of solid nanoparticles as the dispersed phase and liquid as the dispersion medium, a dispersion phase that is the continuous phase in which solid nanoparticles are dispersed.[65] Thus, the term liquid describing the dispersion medium is more strict than fluid.

Application of Maxwell model for calculation of thermal conductivity of ionanofluids with Al_2O_3, Au, metal oxide nanoparticles. Maxwell model was applied for the description of thermal conductivity of ionanofluids with Al_2O_3[11,66] Au,[14] metal oxide nanoparticles (see Figure 5.1c).[9] Prediction of the Maxwell model of thermal conductivity for 1 wt.% Al_2O_3 (γ-phase, $d < 50$ nm) + $[C_4C_1im][NTf_2]$ or $[C_4C_1pyr][NTf_2]$[66] shows lower thermal conductivity enhancement compared to the experimentally measured values (Figure 5.12). The effect of temperature on thermal conductivity is negligible in both experimental and theoretical cases.

Paul et al.[67] also compared thermal conductivity of ionanofluids composed of $[C_4C_1pyr][NTf_2]$ and spherical Al_2O_3 (γ-phase with $d < 50$ nm) or whiskers Al_2O_3 ($d = 2$–6 nm, $l = 200$–400 nm, aspect ratio > 100) at temperatures 283.15–343.15 K. For both types of nanoparticles, the thermal conductivity enhancement shows negligible temperature dependence. The maximum thermal conductivity enhancement of ~9% and ~13% was obtained for 2.5 wt.% of spherical Al_2O_3 and whisker Al_2O_3, respectively. This effect is higher for nanoparticles with a higher aspect ratio,[68] which is in agreement with results obtained for MWCNTs.[12,13] The model values are lower than experimental ones, and differences increase with increasing nanoparticle loading. These differences are larger for ionanofluids with whisker of Al_2O_3, although for the spherical ones, for which the Maxwell model was derived, they are not much smaller (Figure 5.13).[67]

5.5.2 Hamilton and Crosser Model

Hamilton and Crosser[69] modified the Maxwell model for both spherical and non-spherical particles by applying a shape factor n that considers the shape of the

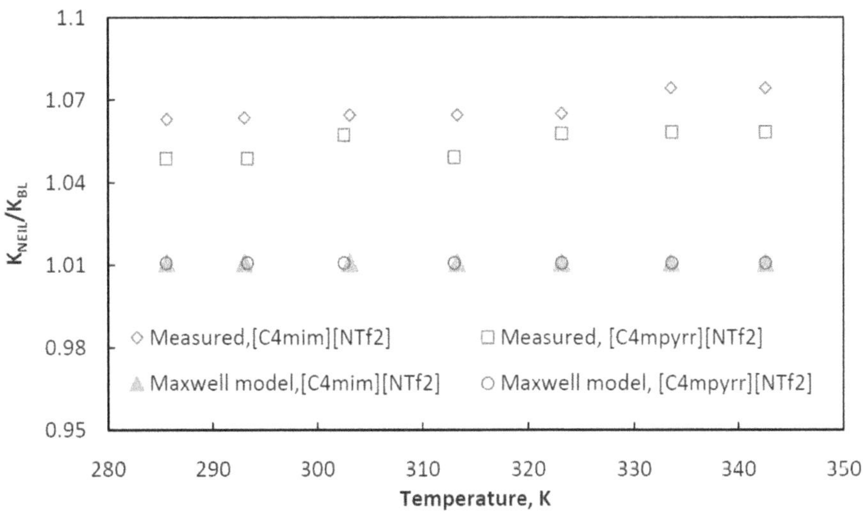

FIGURE 5.12 Normalized effective thermal conductivity of ionanofluids as a function of temperature, thermal conductivity of 1 wt.% Al_2O_3 (γ-phase, d < 50 nm) + $[C_4C_1im][NTf_2]$ or $[C_4C_1pyr][NTf_2]$. (Reprinted with permission from Ref. [66] Copyright © 2013 Elsevier.)

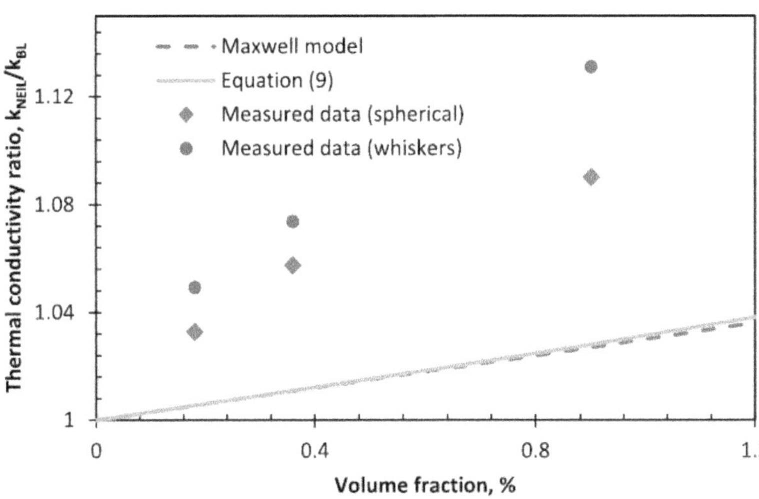

FIGURE 5.13 Thermal conductivity of $[C_4C_1pyr][NTf_2]$-based ionanofluids as a function of volume fraction of spherical and whiskers Al_2O_3, temperature was not specified. The correlation Eq. (9): $\lambda_{INF} = \lambda_{IL}\,(1 + 2.944\varphi_{NP} + 19.672\varphi^2_{NP})$. (Reprinted with permission from Ref. [67] Copyright © 2015 Elsevier.)

dispersed particles. This shape factor is a function of the sphericity, used previously to characterize the shape of particles falling through liquids.[15] Sphericity is defined as the ratio of the surface area of a sphere with volume equal to that of the nanoparticle to the surface area of the nanoparticle. The shape factor, $n \neq 3$ (unlike $n = 3$ for spherical particles), and different nanoparticle shapes were investigated:

$$\lambda_{\mathrm{INF}} = \lambda_{\mathrm{IL}} \frac{\lambda_{\mathrm{NP}} + (n-1)\lambda_{\mathrm{IL}} + (n-1)\varphi_{\mathrm{NP}}(\lambda_{\mathrm{NP}} - \lambda_{\mathrm{IL}})}{\lambda_{\mathrm{NP}} + (n-1)\lambda_{\mathrm{IL}} - \varphi_{\mathrm{NP}}(\lambda_{\mathrm{NP}} - \lambda_{\mathrm{IL}})}. \tag{5.4}$$

The Hamilton and Crosser model shows that for non-spherical particles, a higher increase in thermal conductivity is obtained in comparison to spherical ones with the same volume fraction. Eq. (5.4) reduces to the Maxwell model (Eq. (5.3)) for spherical particles. Moreover, similarly to Maxwell, Hamilton, and Crosser model is static, i.e., it is assumed that particles do not move. The classical models such as Maxwell and Hamilton-Crosser were applied for the description of thermal conductivity of ionanofluids, even though it was previously found to be unable to predict the enhanced thermal conductivity of nanofluids. These models were originally developed for continuum media containing well-dispersed, relatively large particles (the micro- or milli-sized).[70] They did not account for the nanoscale effects such as interfacial nanolayers, clustering, agglomeration, or nanoparticles motion, which played a key role in enhancing the thermal conductivity of nanofluids.

5.5.3 Bruggeman Model

The Bruggeman model is also based on the effective medium theory of nanofluids, but additionally, it considers the clustering of nanoparticles.[6] The thermal conductivity is described by the following relationship:

$$\lambda_{\mathrm{INF}} = \frac{1}{4}\left[(3\varphi_{\mathrm{NP}} - 1)\lambda_{\mathrm{NP}} + (2 - 3\varphi_{\mathrm{NP}})\lambda_{\mathrm{IL}}\right] + \frac{\lambda_{\mathrm{IL}}}{4}\sqrt{\Delta}, \tag{5.5}$$

where

$$\Delta = \left[(3\varphi_{\mathrm{NP}} - 1)^2 \left(\frac{\lambda_{\mathrm{NP}}}{\lambda_{\mathrm{IL}}}\right)^2 + (2 - 3\varphi_{\mathrm{NP}})^2 + 2\left(2 + 9\varphi_{\mathrm{NP}} - 9\varphi_{\mathrm{NP}}^2\right)\left(\frac{\lambda_{\mathrm{NP}}}{\lambda_{\mathrm{IL}}}\right)\right]. \tag{5.6}$$

5.5.4 Aggregation Model

In this model, the aggregation of nanoparticles was considered to describe thermal conductivity.[71] Thus, the Maxwell model was extended with the aggregation factor, the same as for shear viscosity[6]:

$$\lambda_{\mathrm{INF}} = \lambda_{\mathrm{IL}} \frac{\lambda_{\mathrm{a}} + 2\lambda_{\mathrm{IL}} + 2\varphi_{\mathrm{a}}(\lambda_{\mathrm{a}} - \lambda_{\mathrm{IL}})}{\lambda_{\mathrm{a}} + 2\lambda_{\mathrm{IL}} - \varphi_{\mathrm{a}}(\lambda_{\mathrm{a}} - \lambda_{\mathrm{IL}})}, \tag{5.7}$$

where λ_a is the aggregates' thermal conductivity calculated from the correlation equation:

$$\lambda_a = \frac{1}{4}\lambda_{IL}\left\{\left(3\varphi_{in}-1\right)\frac{\lambda_{NP}}{\lambda_{IL}}+\left(3\left(1-\varphi_{in}\right)-1\right)+\left[\left(\left(3\varphi_{in}-1\right)\frac{\lambda_{NP}}{\lambda_{IL}}+\left(3\left(1-\varphi_{in}\right)-1\right)\right)^2+8\frac{\lambda_{NP}}{\lambda_{IL}}\right]^{1/2}\right\}, \quad (5.8)$$

where the $\varphi_{in} = \left(\left(\frac{a_a}{a}\right)^{3-D}\right)$ is the solid aggregate volume fraction. Paul et al.[6] assumed that the aggregation factor $\frac{a_a}{a} = 21.5$, while Asleshirin et al.[72] reported $\frac{a_a}{a} = 16\text{–}18$.

5.5.5 Maxwell–Modified Two-Suffix Margules Gibbs Free Energy (Maxwell–MTSM) Model

Atashrouz et al.[63] treated the difference between the experimental data of thermal conductivity and the values predicted by the Maxwell model as a correction term:

$$\lambda^{m} = \lambda^{real} - \lambda^{Maxwell}, \quad (5.9)$$

where superscripts "m" and "real" correspond to modified property (i.e., correction to Maxwell model) and real experimental data, respectively. The factor λ^m represents the non-ideality of thermal conductivity for ionanofluids. Atashrouz et al. developed a modified two-suffix Margules (MTSM) Gibbs free energy model,[73] which they combined with Eyring's relations[74] to formulate a model for the viscosity of ionic liquid–cosolvent mixtures. Their study showed that the MTSM model effectively captured the behavior of these mixtures when treated as non-ideal systems, specifically for viscosity modeling. Based on this, Atashrouz et al. applied the MTSM model as the non-ideality term in their thermal conductivity model, introducing λ^m as the excess Gibbs free energy G^E contribution:

$$\frac{G^E}{RT} = a_{12}x_{12}x_{21}, \quad (5.10)$$

where α_{12} is an empirical constant and x_{ij} is the local mole fraction of i around j.

Thus, by analogy, λ^m is expressed as follows:

$$\lambda^{m} = \lambda^{E} = \lambda_{12}\phi_{12}\phi_{21}, \quad (5.11)$$

where ϕ_{ij} is the local volume fraction of component i around j and λ_{12} is an empirical constant with thermal conductivity characteristics. Taking into account relations between local volume fractions and bulk volume fractions, the local volume fractions of the molecules in the cells, and the energy G_{ij} between components i and j, the MTSM model was delivered:

$$\lambda^{E} = \lambda_{12}\frac{\varphi_1\varphi_2G_{12}G_{21}}{\left(\varphi_1G_{12}+\varphi_2\right)\left(\varphi_2G_{21}+\varphi_1\right)}, \quad (5.12)$$

where

$$G_{12} = \exp\left(-\frac{\tau_{12}}{RT}\right), \tag{5.13}$$

and

$$G_{21} = \exp\left(-\frac{\tau_{21}}{RT}\right), \tag{5.14}$$

with three energy parameters, namely, λ_{12}, τ_{12}, and τ_{21}, which should be adjusted based on the experimental thermal conductivity data.

Finally, the Maxwell–MTSM model was obtained in the following form:

$$\lambda^{\text{real}} = \lambda_{\text{IL}}\left(1+\frac{3\left(\left(\frac{\lambda_2}{\lambda_1}\right)-1\right)\varphi_1}{\left(\left(\frac{\lambda_2}{\lambda_1}\right)+2\right)-\left(\left(\frac{\lambda_2}{\lambda_1}\right)-1\right)\varphi_1}\right)+\lambda_{12}\frac{\varphi_1\varphi_2 G_{12}G_{21}}{\left(\varphi_1 G_{12}+\varphi_2\right)\left(\varphi_2 G_{21}+\varphi_1\right)}, \tag{5.15}$$

where φ_1 and φ_2 are the volume fractions of the nanoparticle and the ionic liquid, respectively.

5.5.6 Modified Geometric Mean (MGM) Model

Krischer considered an array of elements of specific resistance distributed in a matrix[1] and obtained (Eq. (5.16) was rewritten for ionanofluids):

$$\lambda_{\text{INF}} = \left[\frac{1-f}{\left(1-\varphi_{\text{NP}}\right)\lambda_{\text{IL}}+\varphi_{\text{NP}}\lambda_{\text{NP}}}+f\left(\frac{1-\varphi_{\text{NP}}}{\lambda_{\text{IL}}}+\frac{\varphi_{\text{NP}}}{\lambda_{\text{NP}}}\right)\right]^{-1}, \tag{5.16}$$

where f is an empirical factor equivalent to the fraction of parallel resistances in a rectangular array of elements. This model can be used to determine upper and lower bounds for the thermal conductivity of a heterogeneous system, with $f = 0$ signifying that all particles are arranged in series (creating a high thermal conductivity pathway) and $f = 1$ signifying that all particles are arranged in parallel (creating a low thermal conductivity arrangement). The actual value of f must be obtained by experiment. Hashin and Shtrikman gave more restrictive bounds taking into account the variational theorem[1]

$$\lambda_{\text{IL}}\left[1+\frac{3\varphi_{\text{NP}}\left(\frac{\lambda_{\text{NP}}}{\lambda_{\text{IL}}}-1\right)}{\frac{\lambda_{\text{NP}}}{\lambda_{\text{IL}}}+2-\varphi_{\text{NP}}\left(\frac{\lambda_{\text{NP}}}{\lambda_{\text{IL}}}-1\right)}\right] \le \lambda_{\text{INF}} \le \lambda_{\text{NP}}\left[1-\frac{3\left(\frac{\lambda_{\text{NP}}}{\lambda_{\text{IL}}}-1\right)\left(1-\varphi_{\text{NP}}\right)}{3\frac{\lambda_{\text{NP}}}{\lambda_{\text{IL}}}-\varphi_{\text{NP}}\left(\frac{\lambda_{\text{NP}}}{\lambda_{\text{IL}}}-1\right)}\right]. \tag{5.17}$$

In Eq. (5.17), the lower bound is the Maxwell limit. The upper bound implies a thermal conductivity that is higher than the Maxwell limit when particle aggregation is significant. Turian et al.[64] noted that the upper and lower bounds of the Krischer model are equal to the volume fraction–weighted arithmetic and harmonic means of the thermal conductivities of the two phases. The geometric mean of the two thermal conductivities falls between these bounds and also falls within the more restrictive Hashin–Shtrikman bounds when $\lambda_{NP}/\lambda_{IL} > 5$. Therefore, Turian et al.[64] used the volume fraction–weighted geometric mean of thermal conductivity of the individual phases defined by the following relationship:

$$\lambda_{INF} = \lambda_{IL}\left(\frac{\lambda_{NP}}{\lambda_{IL}}\right)^{\varphi_{NP}}. \tag{5.18}$$

Turian et al.[64] concluded that Eq. (5.18) provides a good estimation of the effective thermal conductivity of suspensions when the ratio of the thermal conductivity of the particle to the thermal conductivity of the base liquid is higher than 3.5. However, in the tested suspensions, particle sizes were relatively large, i.e., of the order of microns.[1] Atashrouz et al.[73] tested the modified geometric mean (mGM) model proposed by Turian et al.[64] for the prediction of thermal conductivity of ionanofluids (Eq. (5.18) was rewritten into ionanofluids).

Comparative study of application of Maxwell, Maxwell–MTSM, and mGM models for calculation of thermal conductivity of ionanofluids. Atashrouz et al.[63] examined Maxwell–MTSM, Maxwell, and mGM models for $[C_2C_1im][N(CN)_2]$ + Baytubes® MWCNTs, $[C_4C_1im][N(CN)_2]$ + Baytubes® MWCNTs, $[C_4C_1pyr][N(CN)_2]$ + Baytubes® MWCNTs, $[C_6C_1im][BF_4]$ + MWCNTs (not specified), $[C_6C_1im][BF_4]$ + graphene (not specified), $[C_4C_1im][NTf_2]$ + Al_2O_3 (γ-phase, $d < 50$ nm), and $[C_4C_1pyr][NTf_2]$ + Al_2O_3 (γ-phase, $d < 50$ nm) ionanofluids. They applied the Maxwell–MTSM model for the description of temperature dependence of thermal conductivity of ionanofluids using the three adjustable energy parameters.[63] The first one is the interaction parameter, assuming symmetry of interactions between the ionic liquid and the nanoparticle. The next two are empirical constants related to the temperature characteristics of thermal conductivity. The mean absolute relative deviation (*MARD*) values are tabulated in Table 5.1 and show good agreement between the model and experimental thermal conductivity. The good agreement of Maxwell–MTSM model correlations with experimental data is also presented in Figures 5.14 and 5.15. However, such good compatibility is not surprising, because as many as three adjustable parameters were used in the Maxwell–MTSM model. Thus, the main disadvantage of this model is the unavailability of experimental data for obtaining energy parameters.

For comparison to Maxwell–MTSM correlation model, Atashrouz et al.[63] applied the mGM and Maxwell models for predicting the thermal conductivity of ionanofluids (see Table 5.1). The values predicted by the mGM and Maxwell models are lower than the experimental ones, but both models reproduce the decrease of thermal conductivity of ionanofluids with increasing temperature (Figure 5.16).

The mGM model displays better accuracy than the Maxwell model with *MARD* below 5.7%, except for two ionanofluids: $[C_6C_1im][BF_4]$ + MWCNTs and $[C_6C_1im]$

TABLE 5.1
The Mean Absolute Relative Deviation (*MARD%*) for Maxwell–MTSM, Maxwell, and mGM Models for Ionanofluids

Ionanofluid	*T* (K)	*MARD (%)* Maxwell–MTSM	Concentration	*MARD (%)* Maxwell	*MARD (%)* mGM	Exp. Data Source
$[C_2C_1im][N(CN)_2]$ + Baytubes® MWCNTs	293.15–343.15	0.68	1 wt.%	–	–	[75]
	293	–	0.3 v/v	3.30	2.31	
	293	–	0.6 v/v	6.54	2.93	
$[C_4C_1im][N(CN)_2]$ + Baytubes® MWCNTs	293.15–343.15	0.59	1 wt.%	–	–	[75]
	293	–	0.29 v/v	6.27	4.38	
	293	–	0.57 v/v	8.83	5.19	
$[C_4C_1pyr][N(CN)_2]$ + Baytubes® MWCNTs	293.15–343.15	0.64	1 wt.%	–	–	[75]
	293	–	0.27 v/v	5.39	3.61	
	293	–	0.55 v/v	7.92	4.35	
$[C_6C_1im][BF_4]$ + MWCNTs	298.15–338.15	1.48	0.03 wt.%	–	–	[76]
	298.15	–	0.018 v/v	5.45	5.03	
	298.15	–	0.037 v/v	14.12	12.15	
$[C_6C_1im][BF_4]$ + graphene	298.15–338.15	0.59	0.06 wt.%	–	–	[76]
	298.15	–	0.018 v/v	5.14	5.03	
	298.15	–	0.037 v/v	12.36	12.14	
$[C_4C_1im][NTf_2] + Al_2O_3$	283.15–343.15	0.18	1 wt.%	–	–	[66]
	Not specified	–	0.36 v/v	5.36	4.45	
$[C_4C_1pyr][NTf_2] + Al_2O_3$	283.15–343.15	0.16	1 wt.%	–	–	[66]
	Not specified	–	0.35 v/v	5.66	5.67	

Source: Reprinted with permission from Atashrouz et al.[63]

$[BF_4]$ + graphene with *MARD* ca. 12% (see Table 5.1). The mGM model has a good performance for prediction without using adjustable parameters.

5.5.7 Murshed, Leong, and Yang (MLY) Model

Tinga et al.[77] found experimentally that, for a given moisture content in wood, the dielectric constant of the cellulose–air–water mixture was significantly higher than would be expected if the water existed as droplets. This effect was explained by the formation of thin layers of water at the cellulose surface. The confocal ellipsoid geometry most closely approximates this effect (see Figure 5.17). Thus, Tinga

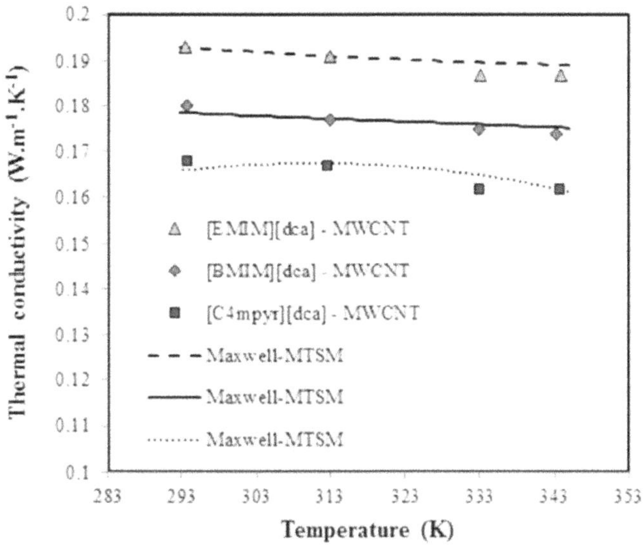

FIGURE 5.14 Comparison between results of thermal conductivity calculated by the Maxwell–MTSM correlation model and experimental data[75] for ionanofluids composed of 1 wt.% Baytubes® MWCNTs and $[C_2C_1im][N(CN)_2]$, $[C_4C_1im][N(CN)_2]$, or $[C_4C_1pyr][N(CN)_2]$. (Reprinted with permission from Atashrouz et al.[63] Copyright © 2013 American Chemical Society.)

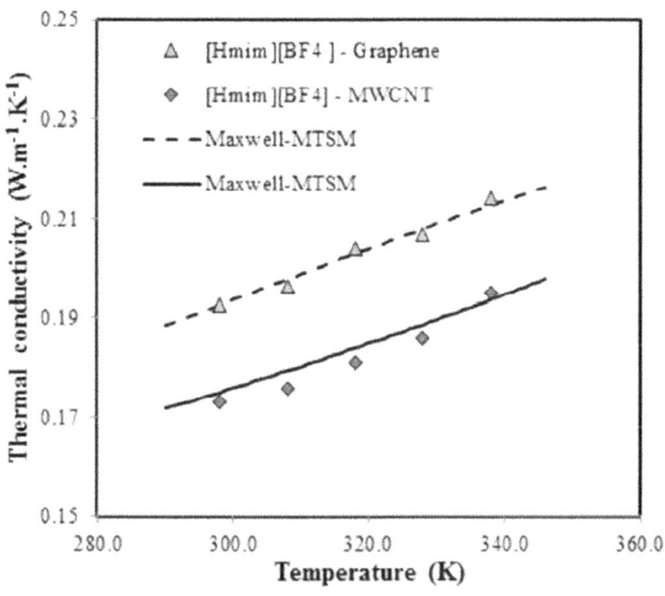

FIGURE 5.15 Comparison between thermal conductivity calculated by Maxwell–MTSM correlation model and experimental data[76] of $[C_6C_1im][BF_4]$ + 0.03 wt.% MWCNTs and $[C_6C_1im][BF_4]$ + 0.06 wt.% graphene. (Reprinted with permission from Atashrouz et al.[63] Copyright © 2013 American Chemical Society.)

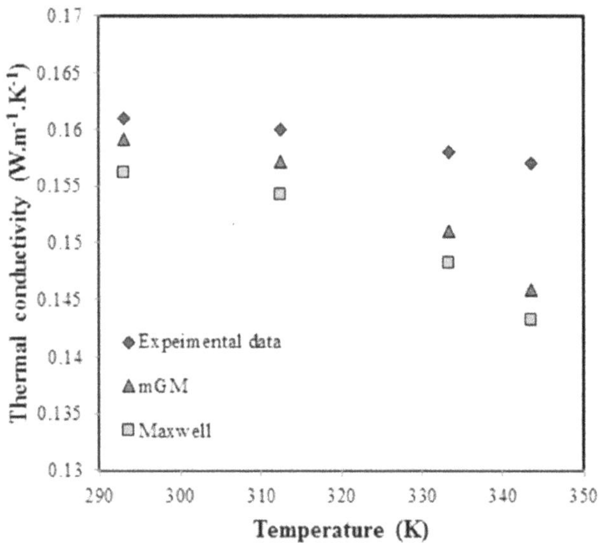

FIGURE 5.16 Prediction of thermal conductivity of $[C_4C_1pyr][N(CN)_2]$ + 0.5 wt.% Baytubes® MWCNTs using Maxwell and mGM models. The experimental data were taken from Ref. [75]. (Reprinted with permission from Atashrouz et al.[63]. Copyright © 2013 American Chemical Society.)

et al.[77] derived a model for the complex dielectric constant of a multiphase mixture (air–water–cellulose) with a confocal ellipsoidal shell (water). Their model for the three-phase spherical inclusions was simplified for thermal conductivity instead of dielectric constant by considering cellulose, water, and air as solid particle, interfacial layer, and host medium, respectively. Based on the structure of the multiphase mixture proposed by Tinga et al.,[77] MLY[25–27] developed a static model for thermal conductivity of nanofluids.

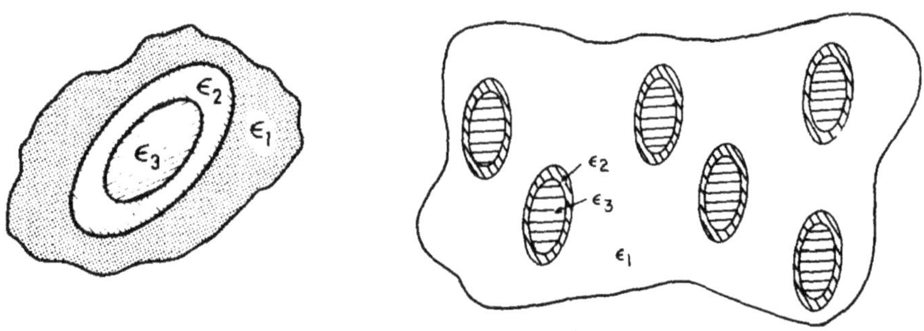

FIGURE 5.17 Model of the confocal ellipsoidal shell and multiphase mixture, where ϵ_1, ϵ_2, ϵ_3 refer to the dielectric constant of air, water, and cellulose, respectively. (Reproduced from Tinga, W. R.; Voss, W. A. G.; Blossey, D. F. Generalized Approach to Multiphase Dielectric Mixture Theory. *J. Appl. Phys.*, **1973**, *44* (9), 3897–3902.[77] with the permission of AIP Publishing.)

This model based on the following assumptions: (*i*) the nanofluid had three components: particle, interfacial layer, and liquid; (*ii*) the particles were spherical and were far apart from each other so that no interactions occur between the particles; (*iii*) particles did not move; (*iv*) the nanotubes of cylindrical geometry had the infinite length; (*v*) the temperature fields were continuous in all three components and at the interfacial boundaries; (*vi*) the temperature gradients were discontinuous at the interfacial boundaries (particle/layer and layer/fluid); the heat fluxes across the interfaces (particle/layer and layer/fluid) and the interfacial boundaries are continuous.[25]

In the proposed formulas, MLY took into account the effects of nanoparticle radius, its concentration, as well as the thickness and thermal conductivity of the interfacial nanolayer. The MLY model was formulated for NFs containing spherical and cylindrical nanoparticles.

For spherical nanoparticles, the thermal conductivity of ionanofluids can be calculated by the MLY model[26] using the following formula:

$$\lambda_{\mathrm{INF(spherical)}} = \lambda_{\mathrm{IL}} \times \frac{\varphi_{\mathrm{NP}} w \lambda_{\mathrm{NP}} + \left(\lambda_{\mathrm{NP}} - w\lambda_{\mathrm{IL}}\right)\left(2g_1^3 - g^3 + 1\right) + \left(\lambda_{\mathrm{NP}} - 2w\lambda_{\mathrm{IL}}\right)2g_1^3\left(g^3(w-1)\varphi_{\mathrm{NP}} + 1\right)}{g_1^3\left(\lambda_{\mathrm{NP}} + 2w\lambda_{\mathrm{IL}}\right) - \varphi_{\mathrm{NP}}\left(\lambda_{\mathrm{NP}} - w\lambda_{\mathrm{IL}}\right)\left(g_1^3 - g^3 - 1\right)}. \quad (5.19)$$

For cylindrical-shaped nanoparticles, the thermal conductivity of ionanofluids can be calculated by the MLY model[26] as follows:

$$\lambda_{\mathrm{INF(cylindrical)}} = \lambda_{\mathrm{IL}} \frac{\varphi_{\mathrm{NP}} w\left(\lambda_{\mathrm{NP}} - w\lambda_{\mathrm{IL}}\right)\left(g_1^2 - g^2 + 1\right) + \left(\lambda_{\mathrm{NP}} + w\lambda_{\mathrm{IL}}\right)g_1^2\left(g^2(w-1)\varphi_{\mathrm{NP}} + 1\right)}{g_1^2\left(\lambda_{\mathrm{NP}} + w\lambda_{\mathrm{IL}}\right) - \varphi_{\mathrm{NP}}\left(\lambda_{\mathrm{NP}} - w\lambda_{\mathrm{IL}}\right)\left(g_1^2 - g^2 - 1\right)}, \quad (5.20)$$

where ω is the ratio of interfacial nanolayer thermal conductivity (λ_{InL}) and thermal conductivity of ionic liquid (λ_{IL}), $\omega = \lambda_{\mathrm{InL}}/\lambda_{\mathrm{IL}}$, γ is the ratio of the thickness of the interfacial nanolayer (h) and radius of the nanoparticle (r) defined as $\gamma = 1 + h/r$ and $\gamma_1 = 1 + h/2r$. The ω depends on the orderliness of liquid molecules in the interface, nature, and surface chemistry of nanoparticles. Leong et al. showed that the predictions were the most reliable for $\omega = 2$, and $h = 1$ nm.[25]

Comparative study of application of Maxwell and MLY models for calculation of thermal conductivity of carbon nanotube-based ionanofluids. The Maxwell model[24] and MLY model[25–27] were used in the first paper of Nieto de Castro et al.[10] for the calculation of thermal conductivity of ionanofluids at 293 K. In the MLY model, Nieto de Castro et al.[10] took into account a difference between the thermophysical properties of pure ionic liquid, nanoparticles, and interfacial nanolayer. Nieto de Castro et al.,[10] and then França et al.,[23] Ribeiro et al.[16] assumed a thickness of an interfacial nanolayer of 2 nm since there was no theoretical model that allowed the estimate of this parameter. As mentioned in Nieto de Castro's work, there is no actual value for λ_{InL} for such systems; hence, this parameter is assumed to be adjustable, with a value between those of the two components of ionanofluid.[10] According to the Baytubes® MWCNTs certificate, an average value of $r = 7.5$ nm, $\rho_{\mathrm{MWCNTs}} =$

185 kg m^{-3}, and λ_{MWCNTs} = 2000 W m^{-1} K^{-1} was assumed. The ρ_{IL} can be obtained experimentally. Based on the above parameters, Ribeiro et al.[16] obtained the thermal conductivity enhancements as a function of the thermal conductivity of the interfacial layer, for [C_4C_1im][NTf_2]. The limiting Maxwell model value is approximately 3ϕ or 0.22,[16] where the thermal conductivity enhancement of the ionanofluid was found to depend only on the thermal conductivity of both phases (solid and liquid) and on the volume fraction of the solid.[2,16] The drawback of the Maxwell model is that it does not consider the basic properties of nanoparticles, such as the interfacial nano-layer and nanoparticle shape.[2] The enhancement of thermal conductivity of [C_4C_1im][NTf_2]-based ionanofluid can be explained by the MLY model with λ_{InL} = 1.4 λ_{IL} and h = 2 nm. However, the obtained results show that for h = 2 nm, the model cannot explain enhancements less than 22%.[10,16] The MLY model can describe the effect of the thermal conductivity of the solid/liquid interface only for systems where the thermal conductivity enhancement is of the order of 20% or greater, as found for [C_4C_1im][NTf_2]-based ionanofluid. For [C_4C_1im][NTf_2]-based INF, Nieto de Castro et al.[10] and Ribeiro et al.[16] suggested that nanosegregation observed earlier for pure ionic liquids also occurs in investigated INFs.[18] However, as the size of the nanotubes is not accounted for in the MLY model, the structure of the ionanofluids is likely to be different compared with that of the base ionic liquid. Nieto de Castro et al.[10] and França et al.[23] calculated thermal conductivity of ionanofluids using the MLY model, assuming that $\lambda_{InL} = \lambda_{IL}$, i.e., the interphase does not play a significant role in the heat transfer process in these systems, at least for small volume fractions. The obtained results showed that the theoretical values are remarkably higher than the experimental ones (except for the lowest volume fraction), and the model does not distinguish between [C_4C_1im][NTf_2] and [C_2C_1im][C_2SO_4] (Figure 5.18).

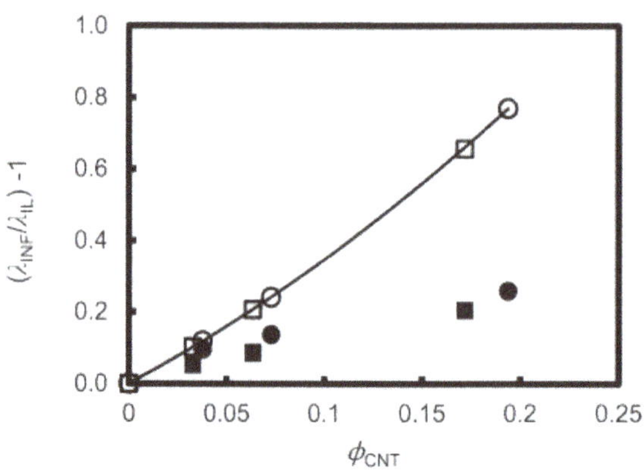

FIGURE 5.18 Thermal conductivity enhancement as a function of the volume fraction of Baytubes® MWCNTs at 293 K, experimental data: ● [C_4C_1im][NTf_2]-based ionanofluids; ■ [C_2C_1im][C_2SO_4]-based ionanofluids; MLY model: ○ [C_4C_1im][NTf_2]-based ionanofluids; □ [C_2C_1im][C_2SO_4]-based[23] ionanofluids. (Reprinted with permission from França, J. M. P.; Vieira, S. I. C.; Lourenço, M. J. V.; Murshed, S. M. S.; Nieto de Castro, C. A. Thermal Conductivity of [C_4mim][$(CF_3SO_2)_2N$] and [C_2mim][$EtSO_4$] and Their INFs with Carbon Nanotubes: Experiment and Theory. *J. Chem. Eng. Data*, **2013**, *58*, 467–476. Copyright 2013 American Chemical Society.)

Therefore, the model, at least with the parameters chosen, is not adequate for volume fractions higher than 4%. The obtained results showed that it was possible to obtain a smaller calculated thermal conductivity of ionanofluids if λ_{InL} is below the range of values assumed by the model, i.e., $\lambda_{IL} < \lambda_{InL} \ll \lambda_{MWCNT}$. The value of λ_{InL} needed to attain the value of experimental thermal conductivity of ionanofluid is inversely proportional to the mass fraction of nanomaterial. This reveals that the more MWCNTs are in suspension, the less relevant the thermal conductivity of the interface of ionanofluid is. The limiting value was obtained for $\lambda_{InL} = 0$ and $h = 0$,[10,16] i.e., no interface is assumed, and it still predicted an enhancement greater than found for all the systems studied.[10] These results are not supported by the TEM and cryo-TEM images, i.e., the interface is much thicker than what was applied by the model theories. Moreover, models do not include the length and size distribution of CNTs.[15] Apart from CNTs having finite length, they curl and are capable of generating additional heat transfer paths that are not included in any of these models.[15] These last ones strongly depend on the type of CNT, i.e., on the nominal length and C-sp^2 to C-sp^3 ratio,[12,13] the C-sp^3-rich CNTs are more curly than the C-sp^2-rich ones.

Comparative study of application of Maxwell, Hamilton and Crosser, and MLY models for calculation of thermal conductivity of carbon nanotube-based ionanofluids. Paredes et al.[15] tested the applicability of Maxwell,[24] Hamilton and Crosser,[69] and the MLY[25–27] models for the estimation of the thermal conductivity of ionanofluids with cylindrical particles. The Hamilton and Crosser model was applied for the Baytubes® MWCNTs with a length of 1 μm, $r = 7.5$ nm, and resulted in a sphericity of 0.323 and therefore the shape factor of 9.28. The MLY model was applied for ionanofluids composed of ionic liquids with the same cation, $[C_4C_1im]^+$, with the same anion, $[NTf_2]^-$, and for the random ionic liquids (Figure 5.10). Paredes et al.[15] took into account for model calculation $h \approx 0.9$ nm suggested by França et al.,[30] and $h = 4$ nm found by Jóźwiak et al.,[13] using cryo-TEM. The next important parameter ω is assumed as an adjustable one,[5,13,15] but Paredes et al.[15] used $\omega \approx 1.2$ obtained from molecular dynamics calculation by França et al.[30] However, application of the model to these two combinations ($\omega = 1.2$; $h = 0.9$ nm) and ($\omega = 1.2$; $h = 4$ nm) cannot reproduce experimental data (Figure 5.19). While the introduction of the MLY model $\omega = 0.5$, i.e., the assumption that $\lambda_{InL} = 0.5\ \lambda_{IL}$, reproduced the experimental data (Figure 5.19). It is worth noting that the Maxwell model aligns well with the MLY model for $\omega = 1.2$ and $h = 4$ nm, whereas the Hamilton and Crosser model significantly overestimates the thermal conductivity enhancement compared with experimental data (Figure 5.19d). It is well known that the addition of nanoparticles causes an increase in thermal conductivity, but the effect of temperature on this change is not clear. In the MLY model, the density and the thermal conductivity of the ionic liquid are temperature-dependent. For $[C_2C_1im][C_2SO_4]$ with 1 wt.% of Baytubes® MWCNTs, experimental data show that the thermal conductivity enhancement rises from 8.6% at 293.8 K to 11.2% at 343.4 K. For 3 wt.%, the enhancement increases from 20.5% to 24.8% over the same temperature range, representing an approximate 25%–30% increase over a 50 K rise.[15] In contrast, the MLY model, with $\omega = 0.5$ and $h = 0.9$ nm, predicts a nearly constant enhancement, independent of temperature, for example, 6.8% at 293.8 K and 6.6% at 343.4 K for 1 wt.%, as well as 20.8% and 20.2%, respectively, for 3 wt.%. These values vary little and fall within the experimental uncertainty.[23]

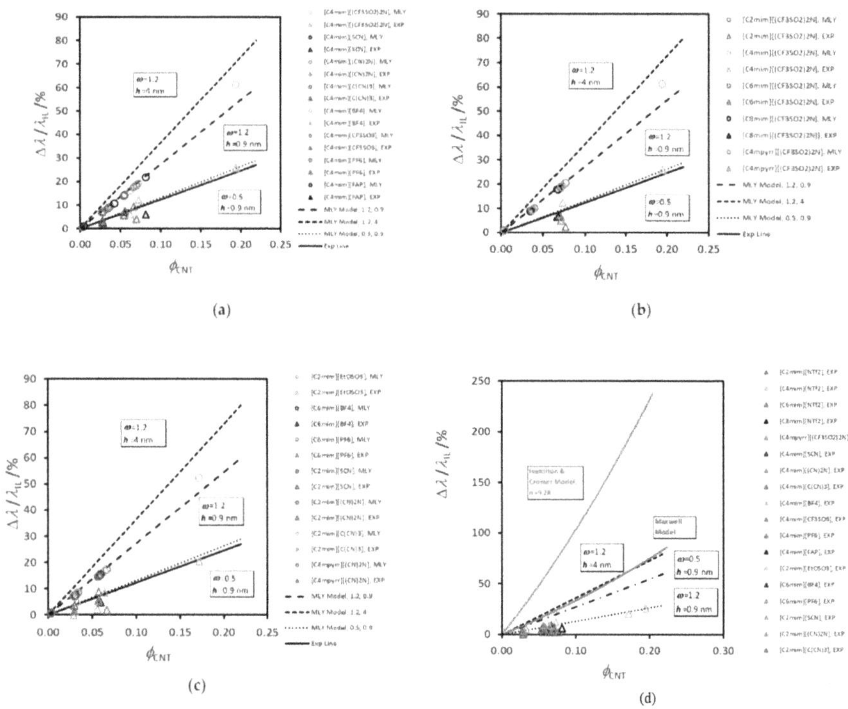

FIGURE 5.19 The effective thermal conductivity of the ionanofluids with MWCNTs at 293 K, experiment and MLY model. (a) Ionic liquids with the same cation; (b) ionic liquids with the same anion; (c) remaining ionic liquids; lines: full line, experimental data trends; — MLY model, (ω = 1.2; h = 4 nm); - - - MLY model, (ω = 1.2; h = 0.9 nm);······· MLY model, (ω = 0.5; h = 0.9 nm); (d) Maxwell, Hamilton, and Crosser and MLY models. Lines: - - - - MLY model, (ω = 1.2; h = 4 nm); -·- MLY model, (ω = 1.2; h = 0.9 nm); ····· MLY model, (ω = 0.5; h = 0.9 nm); — Maxwell model; — Hamilton and Crosser model.[15] (An open-access Creative Commons CC BY license.)

In turn, Nieto de Castro et al.[70] applied the MLY model, with ω = 1.2 and h = 1 nm for predicting the thermal conductivity of $[C_4C_1im][NTf_2]$ + MWCNTs and $[C_2C_1im][C_2SO_4]$ + MWCNTs ionanofluids. Moreover, for comparison, they calculated thermal conductivity using the Hamilton and Crosser model with shape factor n = 6 for cylindrical nanoparticles (Figure 5.20). The MLY model gives a fairly good prediction of the effective thermal conductivity of $[C_4C_1im][NTf_2]$-based ionanofluids at low concentration of MWCNTs. This indicated the poor orderliness or absorption of these ionic liquid molecules on the MWCNT surface. The MWCNT concentration of 3 wt.% in $[C_4C_1im][NTf_2]$ is equivalent to 19.3 vol.%. At such a high concentration, it is anticipated that the increase in thermal conductivity will not be as high as for low concentration, and any classical or recent model can severely over-predict the thermal conductivity at such a high concentration. This is mainly due to the absence of a dynamic mechanism for such a high concentration of nanoparticles. However, the Hamilton and Crosser model over-predicts the thermal conductivity at any concentration.

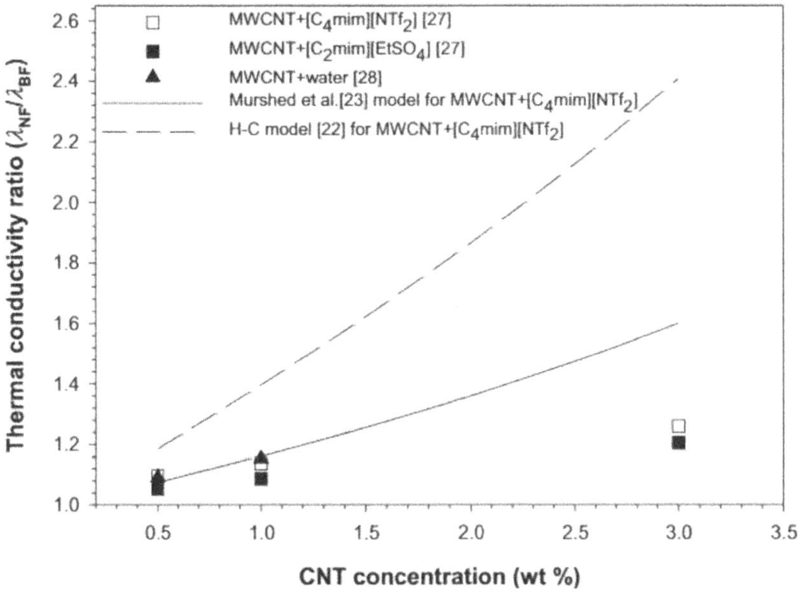

FIGURE 5.20 Comparison of measured and predicted thermal conductivity of nanofluids and ionanofluids as a function of concentration of MWCNT at room temperature. MLY model is the Mursed et al. model, H–C model means the Hamilton and Crosser model. (Reproduced from Nieto de Castro, C. A.; Murshed, S. M. S.; Lourenço, M. J. V.; Santos, F. J. V.; Lopes, M. L. M.; França, J. M. P. Enhanced Thermal Conductivity and Specific Heat Capacity of Carbon Nanotubes Ionanofluids. *Int. J. Therm. Sci.*, **2012**, *62*, 34–39.[70] Copyright © 2012, published by Elsevier Masson SAS. All rights reserved.)

This effect results from a huge difference in thermal conductivity of dispersed MWCNTs and base liquids ($\lambda_{MWCNT}/\lambda_{IL}$ = 16 529), i.e., λ_{MWCNT} = 2000 W m^{-1} K^{-1} and λ[C_4C_1im][NTf_2] = 0.121 W m^{-1} K^{-1}. This classical model might not be valid for such a large thermal conductivity difference. Keblinski et al.[29] showed that λ_{NP} = 10$\lambda_{\text{base liquid}}$ or greater could lead to an increase in thermal conductivity close to the limiting value of thermal conductivity predicted by the Hamilton and Crosser model.

Application of MLY model for calculation of thermal conductivity of carbon nanotube-based ionanofluids. Recently, Jóźwiak et al.[13] found using the MLY model that λ_{InL} was 2.7–45.6 times higher than λ_{IL} with h = 4 nm. The values of the ω parameter show a logarithmic increase with an increasing product of aspect ratio and specific surface area of the dispersed CNTs. Taking into consideration all the above characteristics, the MLY model accurately predicted the thermal conductivity of ionanofluids with 1D carbon-based nanoparticles at 298.15 K.[13] The important fact is that Jóźwiak et al.[13] obtained so far unrivaled thermal conductivity improvement of ~44% and ~68% for ionanofluid and bucky gel composed of [C_2C_1im][SCN] and 1 wt.% in-house 16h MWCNTs (with extreme high aspect ratio of 11,000) and Tuball SWCNTs at 298.15 K, respectively (Figure 5.1). They obtained a weak temperature dependence of thermal conductivity in the range of 293.2–323.2 K. Thus, the MLY model is notably sensitive to the parameters ω and h, which are typically unknown, posing a challenge for practical applications of the model to estimate thermal conductivity enhancement.[15] Nevertheless, if these adjustable parameters can be determined through experiments or molecular simulations, the MLY model may serve as a useful approximate theoretical tool for predicting the thermal conductivity of selected ionanofluids. Supporting this, molecular dynamics simulations by França et al.[30] have confirmed the validity of the MLY model, which assumes the presence of an interfacial nanolayer between CNTs and ionic liquid, where a temperature jump occurs, affecting heat transfer. However, the accuracy of such predictions is ultimately affected by the uncertainty of experimental data.

Comparative study of application of Maxwell, Bruggeman, aggregation, and MLY models for the calculation of thermal conductivity of carbon-based ionanofluids. Asleshirin et al.[72] examined Maxwell, Bruggeman, aggregation, and MLY models for description of thermal conductivity of $[C_6C_1im][PF_6]$ + MWCNTs (d = 90–120 nm, l not specified) and $[C_6C_1im][PF_6]$ + graphene (60–80 nm) ionanofluids. They applied for calculations h = 2 nm and λ_{InL} = 0.35 (reported without units). It is unclear why Asleshirin et al.[72] took into the model calculations r = 15 nm for the examined MWCNTs and graphene. Moreover, MWCNTs and graphene used for experiments were characterized by diameter. In contrast to results reported by Paredes et al.[15] and Nieto de Castro et al.,[70] Asleshirin et al.[72] obtained lower than experimental data and independent of nanoparticle concentration results using the Maxwell and Bruggeman models.

Comparative study of application of Maxwell, Bruggeman, aggregation, and MLY models for the calculation of thermal conductivity of spherical Al_2O_3-based ionanofluids. Paul et al.[71] tested Maxwell, Bruggeman, aggregation, and MLY models for the description of experimental thermal conductivity of ionanofluids composed of $[C_4C_1im][NTf_2]$, $[C_4C_1pyr][NTf_2]$, $[C_4C_1C_1im][NTf_2]$, *N*-butyl-*N*,*N*,*N*-trimethylammonium bis(trifluormethylsulfonyl)imide ([N4111][NTf2]), and spherical Al_2O_3 (γ-phase, d < 50 nm). They noticed no temperature dependence of thermal conductivity for temperatures 283.15–343.15 K. The maximum enhancement of ~11% was observed for $[C_4C_1im][NTf_2]$ + 2.5 wt.% Al_2O_3. Although Bruggeman and aggregation models took into account clustering and aggregation, they still could not predict the experimental thermal conductivity like the Maxwell model. Paul et al.[71] also applied the MLY model, considering $\lambda_{InL} = 3\lambda_{IL}$, h = 2 nm, and r = 20 nm, and they obtained a quantitative agreement with experimental data (Figure 5.21).

The results obtained by Asleshirin et al.[72] for MWCNTs-based ionanofluids who used the same model as Paul et al.[71] are comparable only for the MLY model. The results obtained by the Maxwell model are in line with the results previously reported by Paul et al.[66] and Paul et al.[11] (see Figures 5.12 and 5.13, respectively).

5.5.8 Theory of Timofeeva

Timofeeva et al.[68] clearly indicated that if the particles were spherical, the thermal conductivity of the nanoparticles was much higher than that of the base liquid, and

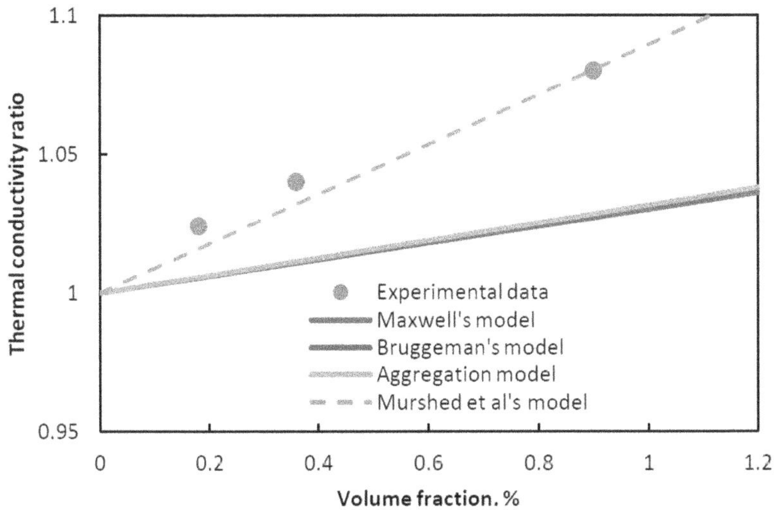

FIGURE 5.21 Thermal conductivity of [C4C1C1im][NTf_2] + Al_2O_3 as function of nanoparticles volume fraction. Temperature was not specified. (Reprinted with permission from Paul et al.[71] Copyright © 2016 Elsevier.)

the volume fraction of nanoparticles was small, the Maxwell Eq. (5.3) could be simplified into the following formula:

$$\lambda_{NF} = \lambda_{IL}\left(1 + 3\varphi_{NP}\right), \quad (5.21)$$

and could be used for the prediction of thermal conductivity of nanofluids.

Comparative study of application of Maxwell, Hamilton and Crosser, spherical MLY, cylindrical MLY, mGM, and Timofeeva models for calculation of thermal conductivity of ionanofluids. Oster et al.[5] examined six of all presented models, while only the mGM model (so-called Atashrouz model) was derived directly for ionic liquid–based nanofluids. Oster et al.[5] tested the these models for ionanofluids composed of [C_4C_1im][$N(CN)_2$], [C_4C_1im][NTf_2], [C_2C_1im][C_2SO_4], [C_4C_1pyr][NTf_2], [C_6C_1im][PF_6],[5] and ionic liquids with trihexyl(tetradecyl)phosphonium cation ([$P_{14,6,6,6}$]$^+$) and acetate, butanoate, hexanoate, octanoate, or decanoate anion,[7] loaded up 3 wt.% of Baytubes® MWCNTs, hexagonal boron nitride (70–80 nm), and graphite (400–1200 nm)[5,7] or mesoporous carbon[7] at temperatures 278–358 K. Oster et al.[5] found that the mGM model gave the most reliable results, because it was based on thermodynamics (modified geometry mean) and included the interactions between ILs and nanoparticles[63] (Figure 5.22).

Oster et al.[5] estimated the influence of nanoparticles' thermal conductivity on the thermal conductivity enhancement in the mGM model (Figure 5.23). The impact of a nanoparticle's thermal conductivity increases logarithmically with increasing concentration, becoming less dependent at high nanoparticle thermal conductivity values. Oster et al.[5] assumed, based on the literature data, that the effect of temperature on the thermal conductivity of nanoparticles was negligible and used the

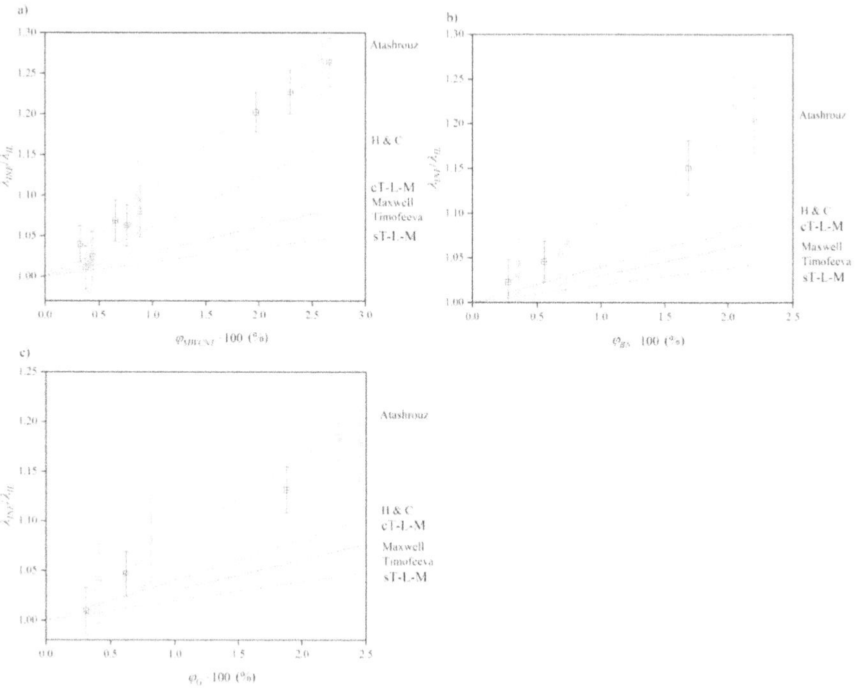

FIGURE 5.22 Thermal conductivity enhancement in comparison to pure ionic liquids as a function of the nanoparticles volume fraction at 298.15 K for (a) MWCNTs-doped ionanofluids; (b) boron nitride-doped (BN) ionanofluids; (c) graphite-doped (G) ionanofluids, for $[C_4C_1im][N(CN)_2]$, $[C_4C_1im][NTf_2]$, $[C_4C_1pyr][NTf_2]$, $[C_2C_1im][C_2SO_4]$, $[C_6C_1im][PF_6]$, including Atashrouz (mGM), Hamilton and Crosser (H–C), cylindrical Tinga–Leong–Murshed (cT–L–M), spherical Tinga–Leong–Murshed (sT–L–M), Maxwell, and Timofeeva models. (Reprinted with permission from Oster.[78])

thermal conductivity of nanoparticles at 298.15 K for the calculations at temperatures up to 358 K using the mGM model. The absolute average relative deviation (*RD*) was found to be 1.44%. The *RDs* between calculated and experimental thermal conductivity of ionanofluids were in the range from +4.19% for $[C_4C_1im][NTf_2]$ + 3 wt.% Baytubes® MWCNTs at 308.15 K to −2.06% for $[C_6C_1im][PF_6]$ + 3 wt.% boron nitride at 298.15 K (Figure 5.24).[5] While for $[P_{14,6,6,6}]$-based ionanofluids, the *RDs* are in the range from 3.65% $[P_{14,6,6,6}]$[HexO] + 3 wt.% boron nitride at 328.15 K to −3.82% for $[P_{14,6,6,6}]$[HexO] + 3 wt.% Baytubes® MWCNTs at 278.15 K.[7] Thus, the mGM model gave reasonable results, taking into consideration the uncertainty of thermal conductivity measurements of 3.44% declared by Oster et al.[5]

The other models, i.e., Maxwell, Hamilton–Crosser, both versions of Tinga–Leong–Murshed (spherical MLY, and cylindrical MLY), and Timofeeva gave significantly lower results than the experimental data (Figure 5.22). Oster et al.[5] explained these results by the assumption of spherical nanoparticles in the case of the Maxwell model, and the fact that the other models did not describe the interactions

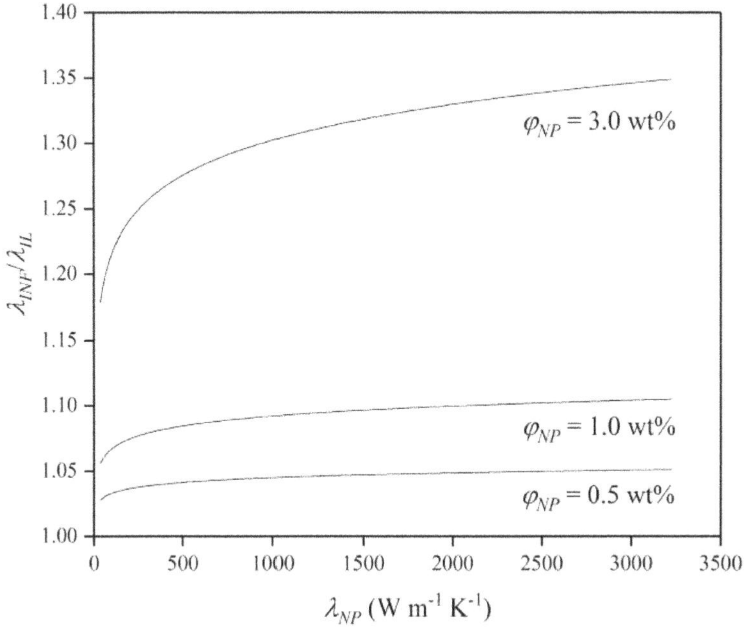

FIGURE 5.23 Impact of nanoparticles' thermal conductivity on the thermal conductivity enhancement for investigated volume concentrations of nanoparticles. Simulation for the mGM model. (Reprinted with permission from Oster.[78])

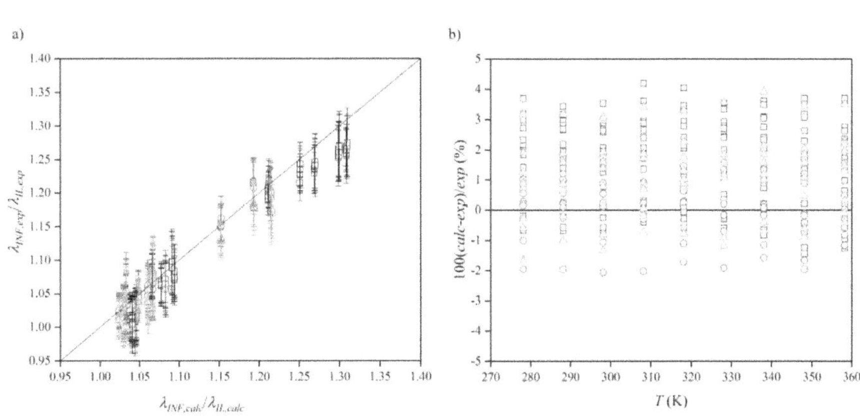

FIGURE 5.24 (a) Thermal conductivity enhancement calculated vs experimental for the whole range of investigated temperature range 278–358 K; (b) deviation between calculated and experimental thermal conductivity enhancement as a function of the temperature. Used model: Atashrouz, for ionanofluid incorporating MWCNT, black squares; boron nitride, red circle; and graphite, green triangles. (Reprinted with permission from Oster.[78])

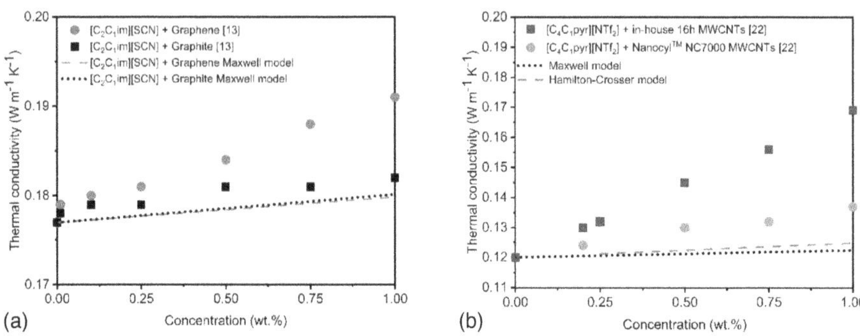

FIGURE 5.25 Thermal conductivity of (a) $[C_2C_1im][SCN]$ + graphene or graphite[13]; (b) $[C_4C_1pyr][NTf_2]$ + in-house 16h MWCNTs or Nanocyl™ NC7000 MWCNTs[22] at 298.15 K. Points – experimental data, dashed, and dotted lines – Maxwell and Hamilton–Crosser models.

between nanoparticles and liquids, which might shift the thermal conductivity to higher values. Oster et al.[5] mentioned that Leong et al.[25] obtained the most reliable results of the MLY model calculations for $\omega = 2$ and $h = 1$ nm. Therefore, it could be assumed that Oster et al.[5] used the same values of ω and h for calculations. In contrast, Paredes et al.[15] obtained higher than experimental data, thermal conductivity predicted by Maxwell, and Hamilton–Crosser models for analogical MWCNTs-based ionanofluids (compare Figure 5.19d and Figure 5.22). Likewise, Nieto de Castro et al.[70] obtained higher than experimental data, thermal conductivity predicted by Hamilton–Crosser model for analogical MWCNTs-based ionanofluids (compare Figures 5.20 and 5.22), while Asleshirin et al.[72] obtained lower than experimental data and independent of nanoparticle concentration results using the Maxwell model. For comparison, Cwynar and Dzida examined the Maxwell model for the description of thermal conductivity of $[C_2C_1im][SCN]$ + 0.1, 0.25, 0.5 wt.% Tuball SWCNTs,[13] $[C_2C_1im][SCN]$ + 0.01, 0.1, 0.25, 0.5, 0.75, 1 wt.% in-house 16h MWCNTs or graphene or graphite,[13] $[C_2C_1im][N(CN)_2]$ or $[C_2C_1im][C(CN)_3]$ + 0.25, 0.50, 0.75, 1.0, 2.0, 3.0 wt.% in-house 16h MWCNTs,[21] $[C_4C_1pyr][NTf_2]$ + 0.5, 1, 2.5 wt.% Al_2O_3.[67] Additionally, they applied the Maxwell model and Hamilton–Crosser with shape factor $n = 6$ for the description of thermal conductivity of $[C_4C_1pyr][NTf_2]$ + 0.2, 0.5, 0.75, 1 wt.% in-house 16h MWCNTs or Nanocyl™ NC7000 MWCNTs[22] at 298.15 K. For all systems, both models underestimated experimental data as can be seen for selected ionanofluids in Figure 5.25.

5.5.9 Brownian Motion – Unexplored Area of Heat Transfer Mechanisms in Ionanofluids

The Brownian motion theory suggests that the enhancement of thermal conductivity is due to the random motion of the nanoparticles in the fluid, which helps to transfer heat more efficiently.[2] Brownian motion corresponds to the random movement of nanoparticles in liquids. This random movement accelerates the nanoparticles collision with fluid molecules and as a consequence, the molecules kinetic energy is converted into thermal energy and hence temperature rises. However, in the discussion of the contribution of the Brownian motion, there is a significant impact due to the increase in temperature. The temperature increment decreases the fluid's viscosity and reduces the agglomeration while increasing the Brownian motion. The Koo and Kleinstreuer model, including Brownian motion effect, was applied for the prediction of thermal conductivity of gold-based ionanofluids by Wang et al.[80] However, the Brownian motion contribution was based on two empirical parameters obtained from experimental data. The Brownian motion and nanoconvection concept have not been explored for ionanofluids so far. Prasher et al.[81] coupled the Maxwell–Garnett conduction model with Brownian-motion-based, convection contribution caused by the Brownian movement of the nanoparticles in their thermal conductivity model for nanofluids. However, this model contains three unknown empirical or fitting parameters.[36] The model of Kumar et al. introduced a model based on Fourier's law of diffusion and the basic idea of kinetic theory to consider the Brownian motion of nanoparticles.[36] However, Murshed and Nieto de Castro[36] pointed out that Kumar et al. postulated that the mean free path of a nanoparticle in fluid was of the order of 1 cm, which was not realistic.

Koo and Kleinstreuer model. The Koo and Kleinstreuer model was derived from the Maxwell model and the experimental data. The equation describing thermal conductivity of ionanofluids was composed of two parts: static, related to Maxwell theory and Brownian:

$$\lambda_{\text{INF}} = \lambda_{\text{static}} + \lambda_{\text{Brownian}} = \lambda_{\text{Maxwell}} + \lambda_{\text{Brownian}}, \tag{5.22}$$

with:

$$\lambda_{\text{Brownian}} = 5 \cdot 10^4 b \varphi_{\text{NP}} r_{\text{IL}} c_{p,\text{IL}} \sqrt{\frac{k_B T}{r_{\text{NP}} d_{\text{NP}}}} f\left(T, \varphi_{\text{NP}}\right), \tag{5.23}$$

where k_B is the Boltzmann constant, T is the nanofluid temperature. To incorporate the influence of temperature as well as interactions between nanoparticles and the surrounding liquid molecules moving around them, two empirical parameters, β and $f(T, \phi_{NP})$, were introduced:

$$b = 0.0137\left(100\varphi_{\text{NP}}\right)^{-0.8229} \tag{5.24}$$

and

$$f\left(T, \varphi_{\text{NP}}\right) = \left(-134.63 + 1722.3\varphi_{\text{NP}}\right) + \left(0.4705 - 6.04\varphi_{\text{NP}}\right)T. \tag{5.25}$$

The relationships (5.24) and (5.25) were determined by utilizing available experimental data.[80] The drawback of the Koo and Kleinstreuer model is the uncertainty of determining the correlation parameters β and $f(T, \varphi_{NP})$. However, this model takes into account the effects of particle size and the dependence of Brownian motion on temperature.

Comparative study of application of Maxwell and Koo and Kleinstreuer models for calculation of thermal conductivity of ionanofluids containing gold nanoparticles.

Wang et al.[80] applied the Koo and Kleinstreuer model to predict the thermal conductivity of the ionanofluids composed of [C_4C_1im][PF_6] with the same volume fraction 0.00102 vol.% of gold nanoparticles but different particle sizes or surface states. The gold nanoparticles were synthesized by a one-phase method (sample 1 with cetyltrimethylammonium bromide, CTAB) or two-phase method (sample 2 with CTAB, sample 3 with poly(*N*-vinylpyrrolidone)40 PVP40, samples 4–5 with 1 wt.% sodium citrate aqueous solutions). The average diameters of the gold nanoparticles in samples 1, 2, 3, 4, and 5 are 5.2, 18.4, 18.4, 29.9, and 56.4 nm, respectively. In calculations, Wang et al.[80] used β and $f(T, \phi_{NP})$ expressed by Eqs. (5.24) and (5.25) without modification. The theoretical predictions of Maxwell and Koo and Kleinstreuer models are both drawn in Figures 5.26 and 5.27. The Maxwell model could reproduce the experimental data well at temperatures lower than 53°C, but underestimates the thermal conductivity enhancements at higher temperatures (Figure 5.26). At lower temperatures (33°C and 43°C), no enhancement in thermal conductivity is observed. This is due to the low nanoparticle concentration, which results in heat transport dominated by the base liquid.[82] However, the thermal conductivity enhancement increases with increasing temperature. Additionally, a peak for thermal conductivity enhancements was found at 18.4 nm (sample 2) for temperatures from 43°C to 70°C (Figure 5.27). This phenomenon is probably related to achieving the high-conductivity percolation paths provided by the synergistic effect of clustering and the collisions among nanoparticles caused by Brownian motion. This interesting phenomenon has also been reported by other researchers[30,83,84]; however, it contradicts most model predictions, including those in Figure 5.26. Additionally, several experimental results suggest that the thermal conductivity enhancement of nanofluids is inversely proportional to nanoparticle size.[7,42–44,85,86] For nanoparticles larger than 18.4 nm, the model qualitatively confirms this effect. As particle size increases, the specific surface area decreases, and particle motion

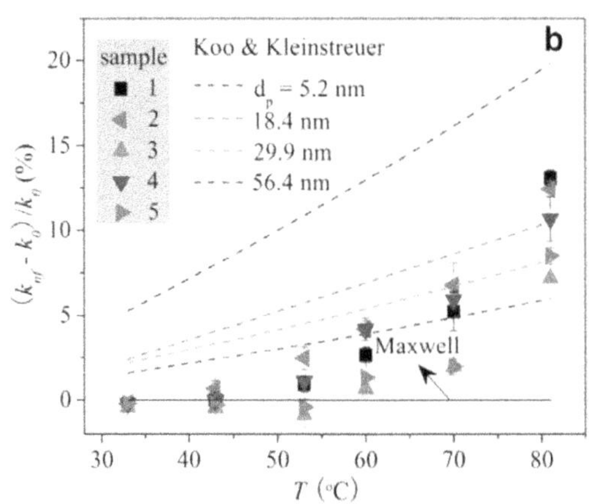

FIGURE 5.26 The thermal conductivity enhancement of samples 1, 2, 3, 4, and 5 as a function of temperature. The theoretical predictions of the Maxwell model (solid line) and the Koo and Kleinstreuer model (dashed lines). (Reprinted with permission from Wang et al.[80] Copyright © 2011 Elsevier.)

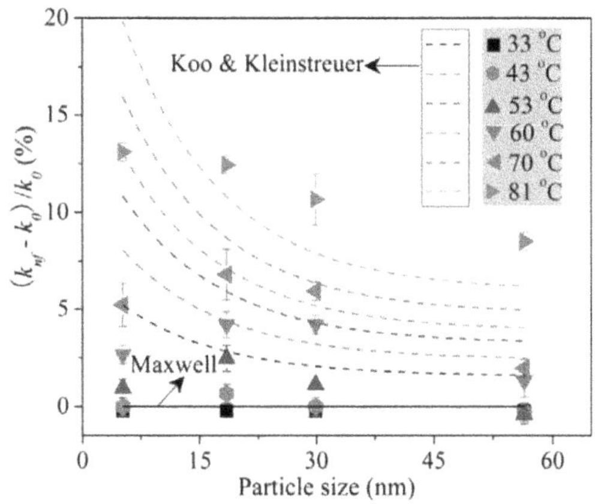

FIGURE 5.27 The thermal conductivity enhancements as a function of particle size at various temperatures for the gold-based ionanofluids (samples 1, 2, 4, and 5) with the same volume fraction of 0.00102%. The theoretical predictions of the Maxwell model (solid line) and the Koo and Kleinstreuer model (dashed lines). (Reprinted with permission from Wang et al.[80] Copyright © 2011 Elsevier.)

slows. Both factors hinder heat transport in nanofluids, causing a gradual decline in thermal conductivity enhancement (see Figure 5.26).

This effect of inverse ratio to the particle size is strengthened at 81°C and becomes consistent with the prediction of the Koo and Kleinstreuer model, which has been observed by others.[87–90]

The ionanofluid composed of the smallest nanoparticles characterizes the highest thermal conductivity enhancement at 81°C (see Figure 5.26). In these conditions, micro-convection induced by Brownian motion begins to dominate the heat transfer process.[31,42] It can be inferred that smaller gold nanoparticles exhibit more intense Brownian motion, which becomes increasingly pronounced with rising temperature. Consequently, the micro-convection of liquid molecules induced by this motion becomes more effective, particularly when the nanoparticles are well dispersed. To sum up, based on the theoretical and experimental investigations, Wang et al.[80] found that the heat transport mechanisms of the gold nanofluids were perplexing, but sometimes correlated with each other. Although the Koo and Kleinstreuer model included Brownian motion and the induced liquid movement, it assumed that nanoparticles were uniformly dispersed in the liquid. As a result, particle clustering and, therefore interactions between particles were not considered. Therefore, the Koo and Kleinstreuer, model could only capture the trend of thermal conductivity enhancement, but it did not accurately match the experimental results. Whatever, the experimental data mostly fall between the predictions of the Maxwell and Koo–Kleinstreuer models, confirming their validity under specific conditions. This also highlights the Brownian motion of gold nanoparticles as a key factor in heat transport, particularly at high temperatures.

The possible heat transport mechanisms in ionanofluids composed of 5.2-nm gold nanoparticles (sample 1) and 18.4-nm gold nanoparticles (sample 2) were presented

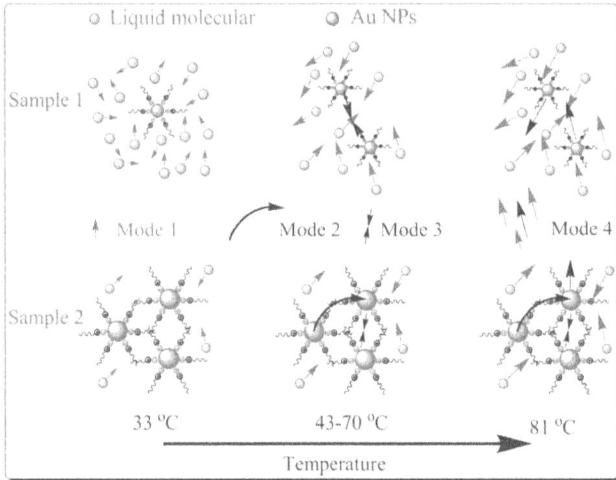

FIGURE 5.28 Schematic illustration of proposed heat transport mechanisms for thermal conductivity enhancements of the gold nanofluids based on experimental clues provided by Wang et al.[80] Modes 1, 2, 3, and 4 stand for the nature heat transport property of base liquid, clustering providing high-conductivity percolation paths, collisions between Au nanoparticles, and micro-convection caused by Brownian motion, respectively. (Reprinted with permission from Wang et al.[80] Copyright © 2011 Elsevier.)

in Figure 5.28. The four modes were introduced, which related to (*i*) the nature of heat transport property of base liquid; (*ii*) clustering providing high-conductivity percolation paths; (*iii*) collisions between gold nanoparticles; and (*iv*) micro-convection caused by Brownian motion, respectively.[31] At temperatures lower than 33°C, the heat transport of ionanofluids with a very low concentration of nanoparticles depends on the nature of the heat transport of the base liquid (mode 1). There is almost no thermal conductivity enhancement (0.00306%), according to the Maxwell model, because the collision of gold nanoparticles induced by Brownian motion is inactive.

A peak value of thermal conductivity enhancement was observed for ionanofluid with optimized clustering and particle size of 18.4 nm (sample 2) in the temperature range from 43°C to 70°C. This system exhibits a synergistic effect of particle collision induced by Brownian motion (mode 2) and clustering (mode 3). For temperatures higher than 81°C, micro-convection of liquid molecules caused by Brownian motion (mode 4) gradually replaces the synergistic effect mentioned above and becomes dominant, although other modes of heat transfer still exist. In summary, sample 1 exhibits the greatest thermal conductivity enhancement at 81°C due to the high temperature, small nanoparticle size, and homogeneous dispersion of gold nanoparticles. All of the above factors favor the induction of micro-convection of liquid molecules. In turn, the ionanofluid composed of CTAB (sample 2) characterizes higher thermal conductivity than ionanofluids composed of PVP40 (sample 3). The longer PVP40 chains, compared with CTAB, increase interfacial thermal resistance and slow gold nanoparticle motion, reducing both synergistic effects and micro-convection. Thus, ionanofluids with bare nanoparticles or coated by short-chain stabilizers can show greater thermal conductivity enhancement. In these conditions, interfacial resistance is reduced, and Brownian motion is intensified. Furthermore, the nanoparticle surface state and the base liquid viscosity affect thermal conductivity enhancement in gold-based ionanofluids by influencing Brownian motion and nanoparticle clustering.

Murshed and Nieto de Castro's concept of Brownian motion in nanofluids. Murshed and Nieto de Castro[35,36] took into account, like in the Koo and Kleinstreuer model, two additive contributions into thermal conductivity of nanofluids, i.e., static and dynamic. The static part has already been developed as Tinga et al.,[77] MLY[25–27] model, taking into account the effects of nanoparticle radius, its concentration, thickness, and thermal conductivity of interfacial nanolayer for nanofluids containing spherical and cylindrical nanoparticles. The dynamic part included the particle motion due to the Derjaguin–Landau–Verwey–Overbeck (DLVO) potential of the interacting complex nanoparticles and Brownian motion potential. In the DLVO theory, the total interparticle potential was the sum of the attractive universal London–van der Waals potential between two spherical particles of the same radius and the repulsive electrostatic potential due to the particle's electric double layer to study the dynamic mechanisms of nanoparticles in the base liquids.[27] The concept of dynamic contribution related to the Brownian motion–based mechanisms of nanoparticles in base liquid was presented in Figure 5.29. Murshed and Nieto de Castro renovated the Brownian motion term (U_{mbm}) by using the effective diffusion coefficient concept, incorporating the effect of nanoparticle concentration:

$$U_{\mathrm{mbm}} = \sqrt{\frac{2k_{\mathrm{B}}T\left(1-1.5\varphi_{\mathrm{NP}}\right)}{m}}, \tag{5.26}$$

where m is the mass of the nanoparticle.

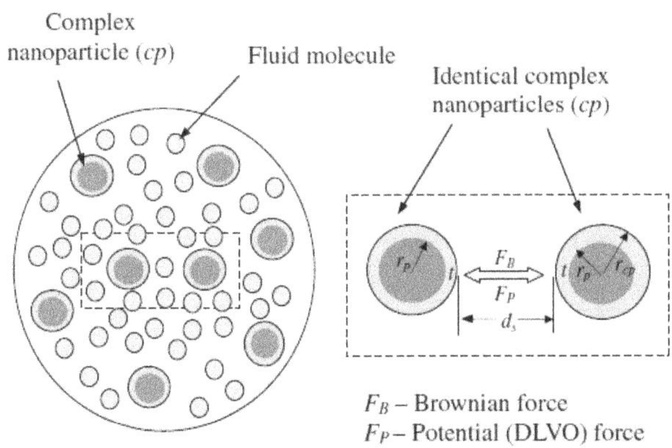

FIGURE 5.29 Conceptual schematic of Brownian motion–based mechanism of nanoparticles in base liquids. InL – interfacial nanolayer termed as cp. (Reprinted with permission from Murshed et al.[27] Copyright © 2011 Elsevier.)

Thus, the higher the particle volume fraction, the weaker the Brownian motion of nanoparticles (see Eq. (5.26)). For uniform complex nanoparticles (nanoparticles with an interfacial nanolayer) in a base liquid (Figure 5.29), the net axial heat flux resulting from the movement of dispersed nanoparticles due to the Brownian force (F_B) impacting on them can be written as follows:

$$q_{\mathrm{bm}} = -0.5\rho_n m_{\mathrm{InL}} c_{\mathrm{NP-InL}} d_s U_{\mathrm{mbm}} \nabla T, \tag{5.27}$$

where ρ_n is the number density, m_{InL}, $c_{\mathrm{NP-InL}}$, and d_s are the mass, specific isobaric heat capacity, and average separation distance of complex nanoparticles, respectively.

Using $\rho_n m_{\mathrm{InL}} = \rho_{\mathrm{InL}}$ in Eq. (5.27), the thermal conductivity of suspensions due to Brownian motion of nanoparticles (λ_{bm}) has the form:

$$\lambda_{\mathrm{bm}} = 0.5 r_{\mathrm{InL}} c_{\mathrm{NP-InL}} d_s U_{\mathrm{mbm}}, \tag{5.28}$$

where ρ_{InL} is the density of complex nanoparticles.

Applying U_{mbm} for complex nanoparticles from Eq. (5.27) into Eq. (5.28) and making use of $\phi_{\mathrm{InL}} = \phi_{\mathrm{NP}}\,\gamma^3$, the following final form of Brownian motion–contributed thermal conductivity is obtained as:

$$\lambda_{\mathrm{bm}} = 0.5 r_{\mathrm{InL}} c_{\mathrm{NP-InL}} d_s \sqrt{\frac{3k_B T\left(1 - 1.5\gamma^3 \varphi_{\mathrm{NP}}\right)}{2\pi r_{\mathrm{InL}} \gamma^3 r_{\mathrm{NP}}^3}}, \tag{5.29}$$

where interparticle separation distance: $d_s = 0.893 r_{\mathrm{InL}} \varphi_{\mathrm{InL}}^{-1/3} = 0.893 r_{\mathrm{NP}} \varphi_{\mathrm{NP}}^{-1/3}$.

These properties of complex nanoparticles are obtained from the formulations given by Murshed et al.[27] The results of the calculation based on Eq. (5.29) shown that the smaller the particle, the larger the Brownian motion–based thermal conductivity λ_{bm} contribution. The Brownian motion contribution to thermal conductivity is significant mainly for smaller-sized particles. The reason is that the smaller the particle size, the greater the movement of particles in the fluid. It is also noted that due to a large interparticle separation distance, the λ_{bm} increases nonlinearly with decreasing particle volume fraction. For each system, the limit value of ϕ_{NP} can be found for which the interparticle separation distance is too large to cause any interaction through the Brownian force of particles. Thus, the Brownian motion contribution to the thermal conductivity under such conditions is not applicable. Compared with existing models, the present one shows significantly better agreement with experimental data. This improvement is attributed to the introduction of both key static and Brownian mechanisms. The results indicate that static mechanisms make the main contribution to thermal conductivity enhancement. Brownian motion, on the other hand, significantly contributes to the improvement of thermal conductivity of nanofluids, with smaller nanoparticles and lower nanoparticle concentrations. Although Murshed and Nieto de Castro have applied the concept of heat transfer by Brownian motion to the prediction of the thermal conductivity of nanofluids with better results. However, they have not yet applied the Brownian motion–induced nanoconvection concept for ionanofluids.

5.5.10 Artificial Neural Network, Machine Learning, AI-Based Tools – Future for Analysis of Thermal Properties of Ionanofluids?

In the last few years, a new trend in predicting thermal properties of ionanofluids has been observed. The classical models described above are increasingly less used for prediction and interpretation of the experimentally obtained results of heat transfer in ionanofluids. Recently, the use of artificial neural networks (ANNs) and machine learning tools has sped up to all engineering fields because of their unique features. Figure 5.30 presents the proposed framework, outlining the process from experimental data to the optimal design of nanofluids aimed at enhancing the efficiency of solar-based renewable energy systems.[91] Now, researchers are using those tools to predict the experimentally obtained data.

An ANNs is a type of soft computational model inspired by the working of neurons in the brain. ANNs make use of concepts of algebra, calculus, statistics, and

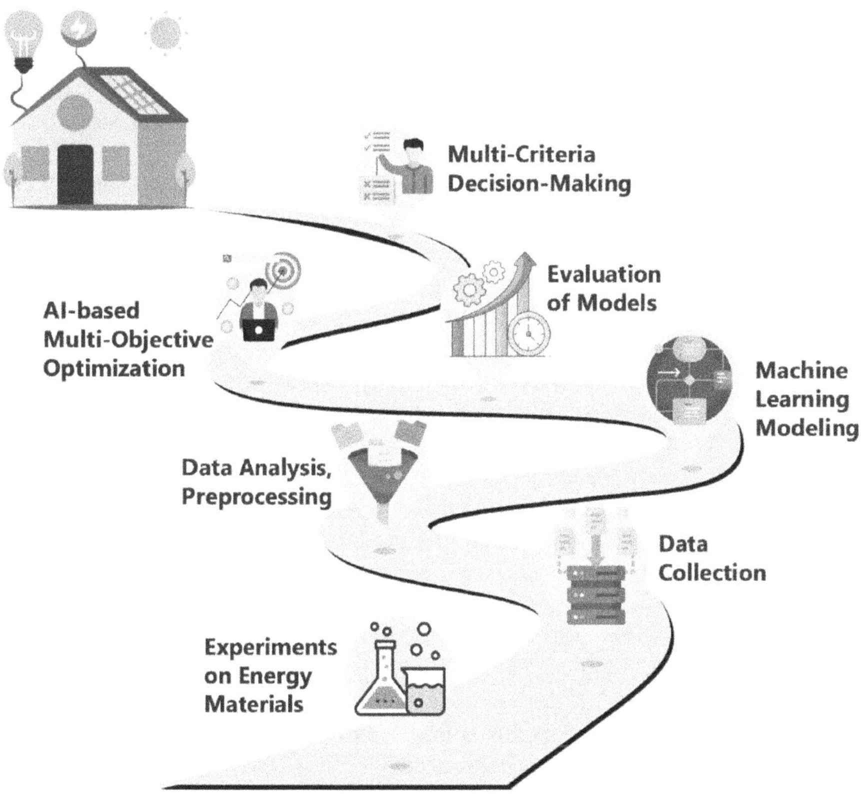

FIGURE 5.30 The proposed framework, outlining the process from experimental data to the optimal design of nanofluids aimed at enhancing the efficiency of solar-based renewable energy systems. (Reprinted with permission from Hai et al.[91] Copyright © 2024 Springer Nature.)

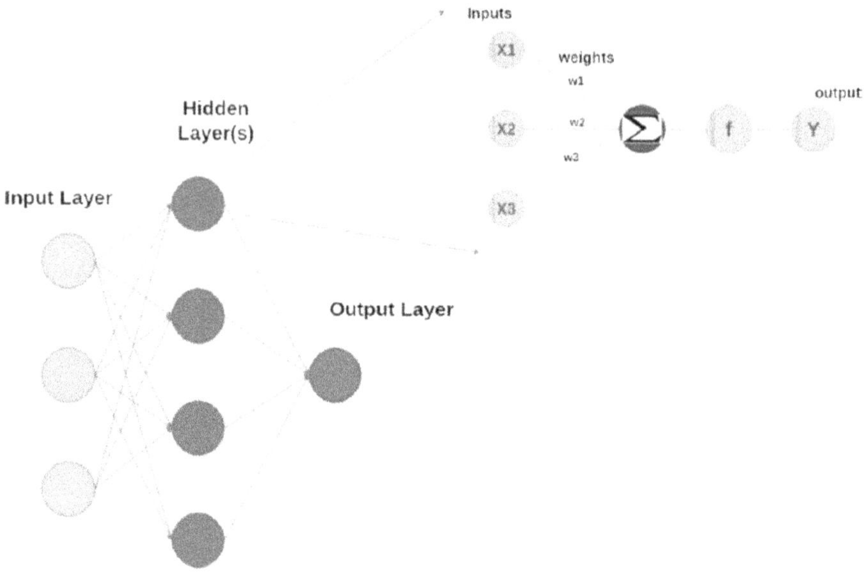

FIGURE 5.31 The structure of a typical ANN model and the working of a neuron. (Reprinted with permission from Parashar et al.[92] Copyright © 2021 Springer Nature.)

probability to learn the mapping between input variables and output variables.[92] The building block of ANNs is known as a neuron, where all the calculations take place. These neurons have some weights and biases associated with them. The structure of a typical ANNs and the neuron's working is shown in Figure 5.31. It shows a structure with three input parameters, four hidden neurons, and one output parameter. In the figure, $X1$, $X2$, and $X3$ are the inputs to the hidden neuron, and $w1$, $w2$, and $w3$ are the weights associated with a hidden neuron. The sigma sign denotes the weighted sum of inputs, f denotes the activation function used, and Y represents the output of a neuron. As ANNs are not programmed to solve a specific problem only, they need to be trained for any particular problem.

The training is a process of finding the values of weights and biases that reduce the error between original values and predicted values to the least possible value. Various algorithms that are used to train the neural networks are gradient descent, Gauss–Newton method, Levenberg–Marquardt, etc.[92] Optimization algorithms such as genetic algorithm, particle swarm optimization, and colony optimization are also commonly used.[92] Parashar et al.[92] developed the ANNs models to predict the thermal conductivity and viscosity of ionanofluids composed of the 0.05, 0.1, and 0.2 wt.% MXene (Ti_3C_2) nanoflakes (size < 100 nm) dispersed in mixture of 1,3-dimethylimidazolium dimethylphosphate [C_1C_1im][DMP] + water (20:80 vol.%) in the temperature range from 19°C to 60°C. The two multiple linear regression models were developed for thermal conductivity. The feedforward network was trained by the Levenberg–Marquardt algorithm. The trial and error method was used to determine the optimum number of neurons in the hidden layer, which was found to be 4. With the help of statistical indices, both the models were found to be well-fit ones; the maximum *RD* was 1.644% for thermal conductivity. Sundar et al.[93] also successfully used an ANNs of Levenberg–Marquardt algorithm for prediction thermal conductivity, viscosity, density, and specific isobaric heat capacity of ionanofluids composed of 0.2, 0.4, 0.6, 0.8, and 1.0 vol.% of nanodiamond/copper oxide (70:30 wt.%) and 1-ethyl-3-methylimidazolium chloride [C_2C_1im]Cl in the temperature range from 30°C to 60°C. Similar research was performed by Shaik et al.,[94] Ben Hamida et al.,[95] Sundar,[96] Jasim et al.,[97] and Sahebalzamani et al.[98]

5.5.11 Reflection on the State of the Art of Thermal Conductivity Mechanisms in Ionanofluids

Despite a large research effort devoted, in the last decade, to identifying the mechanisms and developing theoretical models for heat transfer in ionanofluids, current knowledge on the real heat transfer mechanisms is still scarce.[15,19,37] In 2019, Oster et al. stated that, "*The thickness and thermal conductivity of an interfacial nanolayer are impossible to be measured experimentally...*".[5] While, in 2020, owing to cryo-TEM and TEM imaging, experimental determination of layer thickness was realized by Dzida and Boncel group.[12,13] Thus, the development of experimental methods combined with modeling allows for obtaining an increasingly realistic picture of these complex systems. The currently fashionable machine learning methods need reliable thermophysical and structural results. In this case, critical analysis of experimental data should be a standard.

5.6 FOGGY EFFECT OF NANOPARTICLES ON ISOBARIC HEAT CAPACITY OF IONANOFLUIDS

The second parameter needed for thermal characterization of ionanofluids is isobaric heat capacity. Furthermore, the product of the isobaric specific heat capacity and the density $\left(C_p / V_m = c_p \cdot \rho\right)$ is the energy storage density, which is also called the volumic heat capacity C_p / V_m. The C_p / V_m is the most critical design parameter for thermal fluids because a higher energy storage density value will result in a lower requirement for the thermal fluid volume.[99,100]

5.6.1 Isobaric Heat Capacity of Nanoparticles and Liquids

In contrast to thermal conductivity, the specific isobaric heat capacity of nanoparticles is lower than that of liquids and among them ionic liquids. Nanoparticles do not exhibit specific isobaric heat capacity higher than 1000 J kg^{-1} K^{-1} at 298.15 K. The SWCNTs have a specific isobaric heat capacity of 637 J kg^{-1} K^{-1} (d = 1.25 nm, l-not provided).[101] The MWCNTs, depending on their sizes, have the following specific isobaric heat capacity: 736.1 J kg^{-1} K^{-1} (d = 50 nm, l = 1–2 μm),[102] 770.3 J kg^{-1} K^{-1} (d = 9.5 nm, l = 1.5 μm), and 824.1 J kg^{-1} K^{-1} (d = 60–80 nm, l = 780 μm).[82] For comparison, graphite, fullerene C60, and fullerene C70 in the crystalline phase have an isobaric heat capacity of 709.8, 730.0, and 744.8 J kg^{-1} K^{-1}, respectively.[102] The metallic oxide nanoparticles characterize similar specific isobaric heat capacity, e.g., 692 J kg^{-1} K^{-1} for TiO_2 (d = 50 nm),[103] 670 J kg^{-1} K^{-1} for Fe_3O_4 (d = 23 nm),[103]

745 J kg^{-1} K^{-1} for SiO_2 (d = 30 nm),[104] 765 J kg^{-1} K^{-1} for Al_2O_3 (d = 20 nm),[105] and 536 J kg^{-1} K^{-1} for CuO (d = 20 nm).[105] Metallic nanoparticles have lower specific isobaric heat capacity: 460 J kg^{-1} K^{-1} for Ni (d = 40 nm),[106] 385 J kg^{-1} K^{-1} for Cu (d = 6 nm),[107] 241 J kg^{-1} K^{-1} for Pd (d = 8 nm).[107] In turn, for inorganic liquids, specific isobaric heat capacity varies from 1037 J kg^{-1} K^{-1} for carbon disulfide[108] to 4184 J kg^{-1} K^{-1} for water.[109] Simple, nonpolar organic compounds such as heptane have a specific isobaric heat capacity of 2157 J kg^{-1} K^{-1}[110]; on the other hand, carbon tetrachloride has a specific isobaric heat capacity of 854.0 J kg^{-1} K^{-1}.[111] The polar organic compounds such as 1-butanol and 1,2-ethanediol have specific isobaric heat capacity of 2389 J kg^{-1} K^{-1}[112] and 2469 J kg^{-1} K^{-1},[113] respectively. The specific isobaric heat capacity of aprotic ionic liquids varies from 1281 J kg^{-1} K^{-1} for $[C_1C_2im][NTf_2]$,[82] through a moderate value of 1684 J kg^{-1} K^{-1} for $[C_1C_2im][SCN]$,[82] up to 2022 J kg^{-1} K^{-1} for $[P_{14,6,6,6}][OAc]$.[7] As for protic ionic liquids, the specific isobaric heat capacity can attain values up to 3625 J kg^{-1} K^{-1} for imidazolium octanoate.[114] Based on the classical mixing rules, the resulting isobaric heat capacity of a mixture is determined by the contributions of its individual components. In the case of ionanofluids, the solid nanoparticles generally exhibit a significantly lower isobaric heat capacity compared with that of the ionic liquids. Moreover, these solids are usually present only in relatively small amounts. On the other hand, the liquid phase, represented by the ionic liquid, has a higher isobaric heat capacity than the solid phase. Consequently, when solids are dispersed within the ionic liquid, the overall isobaric heat capacity of the ionanofluid is expected to be lower than that of the pure ionic liquid. Therefore, the experimentally observed enhancement in the isobaric heat capacity of ionanofluids should be regarded carefully.

5.6.2 Specific Isobaric Heat Capacity of Ionanofluids

A critical analysis of available literature data of the specific isobaric heat capacity of ionanofluids showed discrepancies. For example, Nieto de Castro et al.[10] achieved a maximum increase of 8% in the specific isobaric heat capacity at ~363 K, in comparison to ionic liquid IL, for $[C_4C_1im][PF_6]$ + 1 wt.% Baytubes® MWCNTs. In turn, Shevelvoya et al.[115] for the analogous system (MWCNTs, d = 10–80 nm, l = 1–2 μm), obtained 0.3% decrease in the specific isobaric heat capacity at 298 K. Similarly, contradictory results were observed for $[C_6C_1im][PF_6]$ + 1 wt.% Baytubes® MWCNTs or MWCNTs (d = 20–30 nm, l = 10–30 μm).[116,117] Oster et al. reported 9% increase in the specific isobaric heat capacity of ionanofluids at ~298.15 K, while Boldoo et al. observed ~10% decrease in the specific isobaric heat capacity at 303 K with a similar concentration of MWCNTs. One more contradiction can be found for $[C_4C_1im][NTf_2]$ + 1 wt.% MWCNTs.[85,116] Oster et al. reported ~7% increase in the specific isobaric heat capacity at 298.15 K, whereas Parmar et al. obtained ~0.5% decrease in the specific isobaric heat capacity at 298.15 K. Boldoo et al. investigated several $[C_6C_1im]$-based ionanofluids and found that the addition of MWCNTs to ionic liquids led to a decrease in the specific isobaric heat capacity of ionanofluids. Regardless of the ionic liquid type, with an increasing concentration of MWCNTs, the specific isobaric heat capacity of ionanofluids decreased by up to 15% at temperatures from 303 to 383 K. In contrast, Oster et al.[7,116] investigated several ionanofluids with Baytubes® MWCNTs, and obtained an increase in the specific isobaric heat capacity, varied from 5% to 15%, for different ionic liquids and loadings of MWCNTs at temperatures 293–363 K. Wang et al.[76] observed a decrease of 1% in the specific isobaric heat capacity for $[C_6C_1im][BF_4]$ + 0.06 wt.% MWCNTs at 298.15 K. Parmar et al.[85] found that carbon chain length plays a significant role in the mechanism of changing the specific isobaric heat capacity of ionanofluids due to the addition of MWCNTs ($d = 6–9$ nm, $l = 5$ μm). For $[C_2C_1im][NTf_2]$ + 1 wt.%, MWCNTs decrease in the specific isobaric heat capacity of ionanofluids was observed, while, for $[C_{12}C_1im][NTf_2]$ + 1 wt.%, MWCNTs increase. While Oster et al.[7] found no dependence of the anion chain length on the isobaric heat capacity enhancement, Jóźwiak et al.[13] obtained a slight decrease in the specific isobaric heat capacity of $[C_2C_1im][SCN]$ + in-house 16h MWCNTs, but of order measurement uncertainty. Cwynar et al.[82] showed that for 43 ionanofluids composed of different ionic liquids and various types of MWCNTs, the effect of addition of MWCNTs did not exceeded 3% at temperatures 293.15–323.15 K. Reported by Paul et al.,[71] 62% the specific isobaric heat capacity increase for $[C_4C_1pyr][NTf_2]$ + 2.5 wt.% Al_2O_3 at 303.15 K was unreliable, due to the unreliable specific isobaric heat capacity of pure ionic liquid. Thus, it is extremely difficult to assess the nature of CNTs influence on the specific isobaric heat capacity of ionanofluids due to outright contradictory experimental results published up to date, resulting among others from the specific isobaric heat capacity measurement difficulties.

5.6.3 Theories of Isobaric Heat Capacity Enhancement for Ionanofluids

Ideal gas theory. The mechanism of the isobaric heat capacity enhancement has been widely discussed based on the mixing theory for ideal gas mixtures. However, it has been shown that this approach is not relevant for nanofluids, where the specific isobaric heat capacity of nanoparticles is lower than the specific isobaric heat capacity of liquids. Therefore, the enhancement of isobaric heat capacity cannot be observed.[13] Moreover, as in the case of thermal conductivity, from the thermodynamic point of view, it is not correct to add up quantities of substances in different phases. In this case it is not correct to add up the specific isobaric heat capacity of a solid nanoparticle and the specific isobaric heat capacity of an ionic liquid, which is in the liquid state (see Eq. (5.16)). In the case of solutions with a solid solute, a non-symmetric reference system is used. In the case of nanofluids, it is a more complicated issue, because the final system is composed of solid and liquid phases. Therefore, the simple mixing rules are not suitable here.

Intermolecular interaction mechanisms for explanation of the specific isobaric heat capacity enhancement in nanofluids. The attempt to explain the specific isobaric heat capacity enhancement for nanofluids with respect to the base liquid was proposed by Shin and Banerjee.[84] Shin and Banerjee[84] proposed three mechanisms based on: (*i*) higher specific isobaric heat capacity of nanoparticles, than the bulk phase, due to higher specific surface energy; (*ii*) solid–liquid interaction energy, i.e., additional thermal storage mechanisms due to interfacial interactions (e.g., such as interfacial thermal resistance and capacitance) between nanoparticle and the adhering liquid molecules due to the extremely high specific surface area of the nanoparticle; (*iii*) "layering" of liquid molecules at surface to form semi-solid layer adhering to the nanoparticles, which are likely to have enhanced specific heat capacity due to the smaller intermolecular spacing similar to the nanoparticle lattice structure on the surface compared with the higher intermolecular spacing in the bulk liquid phase (see Figure 5.32). Shin and Banerjee noticed that the specific isobaric heat capacity of Al_2O_3 nanoparticles can be enhanced by up to 25% compared with the bulk phase due to their high surface energy per unit mass. Wang et al.[31] demonstrated theoretically that decreasing nanoparticle size increases their specific isobaric heat capacity. This implies that models should take into account the dependence of the specific isobaric heat capacity of nanofluids on nominal nanoparticle diameter. The underlying mechanism arises from the behavior of surface atoms, which, due to fewer bonds, are less constrained. These atoms vibrate at lower natural frequencies and higher amplitudes, contributing to higher surface energy. Consequently, the phonon spectrum of nanoparticles becomes quantized and size-dependent, reflecting discrete vibrational states constrained by particle dimensions (see Mode I in Figure 5.32). Moreover, the extremely high surface area per unit mass of nanoparticles leads to a significant increase in interfacial thermal resistance with surrounding liquid molecules. This high interfacial thermal resistance contributes to additional thermal storage, arising from vibrational energy interactions between nanoparticle atoms and interfacial molecules (see Mode II in Figure 5.32). Similarly to thermal conductivity, the concept of a semi-solid layer of liquid molecules at the surface of a nanoparticle was applied for the explanation of the specific isobaric heat capacity enhancement of nanofluids (see Mode III in Figure 5.32). This "adhesion layer" of ~2–5 nm in thickness, which is about 10 molecules or less in thickness, would depend on the surface energy of the nanoparticle. These semi-solid layers usually have higher thermal properties than the bulk liquid and therefore contribute to increasing the effective specific isobaric heat capacity of nanofluid. The number of adhered liquid layers depends on the surface energy of the crystalline interface. As nanoparticle size decreases, the mass fraction of these "semi-solid" layers increases for the same weight concentration of particles. This suggests the existence of an optimal nanoparticle diameter that maximizes the amount of adhered molecules and, in turn, the enhancement of the nanofluid's specific isobaric heat capacity. The optimal nanoparticle concentration may likewise depend on nominal nanoparticle size or size distribution.

In the case of ionanofluids, Oster et al.[7,116] were the first to demonstrate that the experimentally observed enhancement in isobaric heat capacity can be explained by the presence of interfacial nanolayering on the nanoparticle surface. Moreover, as pointed out by Hentschke,[86] the isobaric heat capacity enhancement depends on the nanoparticle size (in reverse order for the surface area), which is basically in a good agreement with experimental data obtained by Oster et al.[7,116] Oster et al.[7,116] observed that the intensity of isobaric heat capacity enhancement changed as follows: graphite > mesoporous carbon > boron nitride > carbon nanotubes. However, in case of mesoporous carbon, the enhancement is higher than for boron nitride but lower than graphite. This change may be the result of the very high porosity of mesoporous carbon (average pore size 20–40 nm), which influences the adsorption of liquid molecules and, therefore, the enhancement. Oster et al.[7,116] stated that the temperature dependence of the isobaric heat capacity enhancements was affected by the nanoparticles size. They found that for carbon nanotubes-based ionanofluids, the

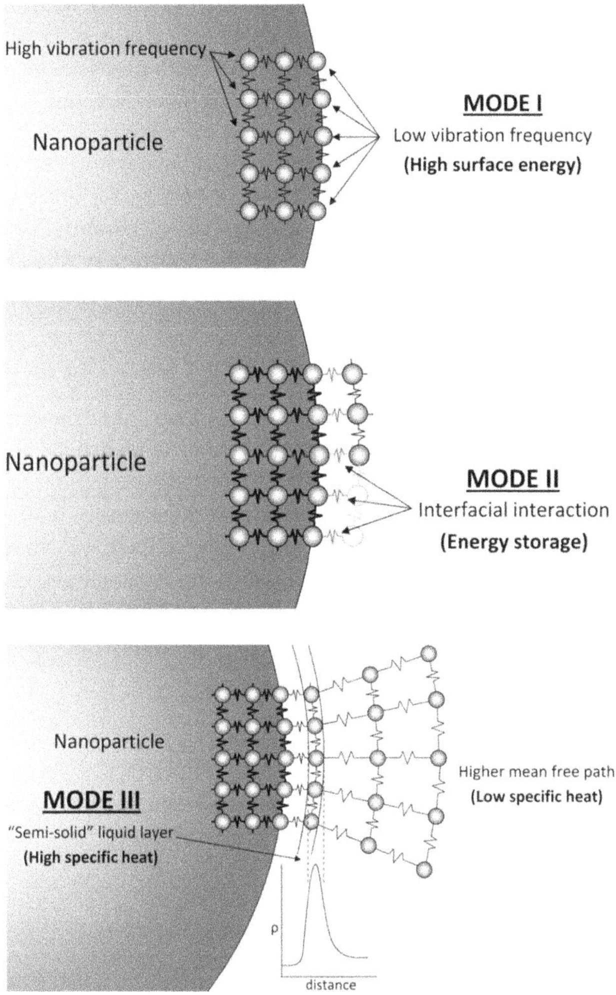

FIGURE 5.32 Schematic showing of the I, II, III modes. (Reprinted with permission from Shin and Banerjee.[84] Copyright © 2020 Elsevier.)

Mode I of energy storage arising from high surface energy of the surface atoms in the lattice (low vibration frequency and higher amplitudes of vibration leading to higher surface energy). The bonds "visualized" to act like springs.

Mode II of energy storage arising from interfacial interactions between liquid molecules and lattice atoms – which act as virtual spring-mass systems.

Mode III of energy storage arising from a layer of semi-solid molecules adhering to the surface lattice atoms. The layer of semi-sold molecules is expected to have a higher specific isobaric heat capacity due to the reduced intermolecular spacing and semi-solid behavior of this layer. Such a layer of semi-solid molecules would not exist in the absence of the nano-particles (key: ρ = mass density; units: kg/m^3).

isobaric heat capacity enhancement decreased with increasing temperature, which might be related to an increase of the hydrophobicity, i.e., less favored attraction of the liquid molecules. In the case of ionanofluids with graphite, the isobaric heat capacity enhancement increased with increasing temperature due to a hydrophobicity decrease, which leads to easier creation of nanolayers. Mesoporous carbon characterizes relatively high hydrophobicity at room temperature, similar to graphite, and therefore displays a comparable behavior. While, for boron-nitride-based ionanofluids, the isobaric heat capacity enhancement was approximately independent on the temperature due to the diversely charged surface of charge distribution between nitrogen and boron.

5.6.4 Mathematical Models of Isobaric Heat Capacity of Ionanofluids

Aside from the experimental investigation of ionanofluids, some theoretical attempts were made to explain the effects of the addition of the nanoparticles to the base liquids on the specific isobaric heat capacity,[83,84,86] and some equations to estimate the specific isobaric heat capacity of the nanofluids were proposed. It should be noted that all these works were based on molecular liquids, so there was an inherent inaccuracy in employing them to the determination of the ionanofluids' characteristics. There are two main ways of calculating the specific isobaric heat capacity of nanofluids in this case applied to ionanofluids. The first equation stems from the rules of mixing from ideal gas theory:

$$c_{p,\mathrm{INF}} = \varphi c_{p,\mathrm{NP}} + (1-\varphi)c_{p,\mathrm{IL}}. \tag{5.30}$$

Zhou et al.[83] showed that the following relationship more accurately corresponded to the effective isobaric specific heat c_p of nanofluids:

$$c_{p,\mathrm{INF}} = \frac{\varphi(\rho c_p)_{\mathrm{NP}} + (1-\varphi)(\rho c_p)_{\mathrm{IL}}}{\varphi\rho_{\mathrm{NP}} + (1-\varphi)\rho_{\mathrm{IL}}}. \tag{5.31}$$

Eq. (5.31) was rewritten to ionanofluids.

Eq. (5.31) can be rewritten to mass fraction, w, of nanoparticles ~~NPs~~:

$$c_{p,\mathrm{INF}} = wc_{p,\mathrm{NP}} + (1-w)c_{p,\mathrm{IL}}. \tag{5.32}$$

The mechanism of the isobaric heat capacity enhancement for ionanofluids based on ionic liquids nanolayer on the nanoparticles surface, described by Shin and Banerjee[84] was further extended by Hentschke[86] who discussed the possible mechanism of the isobaric heat capacity enhancement of nanofluids in relations to base liquid. It was assumed that the isobaric heat capacity enhancement may be caused by one or both of the following factors: (*i*) existing of the nanolayer composed of liquid molecules with the higher isobaric heat capacity in comparison to base liquid and dependent on the type of nanoparticles; (*ii*) existing of the mesolayer interacting with other mesolayers with the assumption that the enhancement occurs until some specific loading of nanoparticles, and then becomes a decreasing function of nanoparticles concentration. Nevertheless, all of these theories are based on molecular solvents-based nanofluids, and none of them considered ionanofluids (the maximum mass fraction concentration was always below 3%). The mesolayer is a significantly wider interfacial layer induced by a nanoparticle in the surrounding fluid. However, no experimental results have indicated the existence of interacting mesolayers. Hentscke[86] described the specific isobaric heat capacity of the nanofluid considering the impact of nanolayer (relation was rewritten into ionanofluids):

$$c_{p,\mathrm{INF}} = \frac{M_{\mathrm{IL}} c_{p,\mathrm{IL}} + M_{\mathrm{InL}} c_{p,\mathrm{InL}} + M_{\mathrm{NP}} c_{p,\mathrm{NP}}}{M_{\mathrm{IL}} + M_{\mathrm{InL}} + M_{\mathrm{NP}}}. \tag{5.33}$$

Hentscke, assumed that the h and $c_{p,\mathrm{InL}}$ were adjustable parameters, and $\rho_{\mathrm{InL}} = \rho_{IL}$. Cwynar et al. used h determined by cryo-TEM,[12,13] while ρ_{InL} and $c_{p,\mathrm{InL}}$ was estimated, similarly to Eq. (5.33).[82] The density of interface is closely corelated with density of bulk liquid, thus $\rho_{\mathrm{InL}} = \rho_{IL} D$, where D is constant by which density of interface differs from bulk liquid. The c_p of interface was treated in the same way $c_{p,\mathrm{InL}} = c_{p,\mathrm{IL}} K$, where K is constant by which the specific isobaric heat capacity c_p of the interface differs from ionic liquid. Finally, the following relationship was obtained:

$$c_{p,\mathrm{nf}} = w_{\mathrm{np}} c_{p,\mathrm{np}} + w_{\mathrm{f}} c_{p,\mathrm{f}} + \rho_{\mathrm{f}} D A \Delta w_{\mathrm{np}} c_{p,\mathrm{f}} \left(K - 1\right), \tag{5.34}$$

where K and D are adjustable parameters.

According to Eq. (5.34), when $D > 1$ the specific isobaric heat capacity of interface is greater, which was indirectly proposed by Shin and Bannerje.[84] With higher density of semi-solid layer, the lower mean free path of particles results in a higher specific isobaric heat capacity of interface. It is only reasonable that density of adsorbed phase is greater than of bulk phase. Also, the higher specific isobaric heat capacity of interface could be linked with the interaction between bulk ionic liquid and adsorbed phase, which was also proposed by Shin and Bannerje. Cwynar et al. assumed that thickness and density are the same at every point of the nanolayer. Calculations were made for $[C_4C_1pyr][NTf_2]$ + 0.20 or 1.0 wt.% Nanocyl™ NC7000 MWCNTs and $[C_4C_1pyr][NTf_2]$ + 0.25 or 1.0 wt.% in-house 16h MWCNTs, for which all experimental data were available at 298.15 K.[82] Constant D and K were adjusted to experimental data, but the value of $D > 1.5$ and $K > 4$ were not reasonable. The best agreement between theoretical and experimental data, no higher than measurement uncertainty (±2%), was obtained for $D = 1.1$ and $K = 1.1$. Such small values of D and K suggest that interface differs not so much from bulk ionic liquid. Thus, the nanolayer has a little or none influence on the specific isobaric heat capacity of ionanofluids. A new model needs to be developed for explanation of any significant change in the specific isobaric heat capacity.

5.6.5 Group Method of Data Handling Model for Prediction of Isobaric Heat Capacity of Ionanofluids

Sadi applied the group method of data handling technique to predict isobaric heat capacity of ionanofluids as a function of reduced temperature, acentric factor, molecular weight of pure ionic liquids, and concentration of nanoparticles.[117] In this case, the reliable experimental data are crucial for training and constructing a truthful model. Unfortunately, the selected experimental data are not reliable, even for pure ionic liquids [author's note]. Therefore, only the general idea of these calculations will be presented here. Sadi divided the collected experimental data (528 data points) into training (70%) and testing (30%) subsets.[117] The training set was used to predict model coefficients and train the optimal network, while the testing set was used for model validation.

The group method of data handling identifies polynomial models directly from experimental data. It uses feed-forward networks with short polynomial transfer functions to link inputs to outputs, while the polynomial coefficients are determined through regression combined with self-organizing learning principles. Model construction proceeds iteratively. Each iteration trains the network, computes the unknown coefficients, and selects the most accurate expressions. A new hidden layer is added at each step, choosing the best combinations of input variables. The final structure consists of quadratic polynomials. To ensure optimal network configuration, nodes in each hidden layer can use any independent variables or nodes from previous layers, as long as the polynomial order remains below three. This crossover among layers increases model complexity and the range of input combinations, improving accuracy when modeling nonlinear systems. The structure of developed group method of data handling model is schematically shown in Figure 5.33. As can be seen, the proposed model has four parameters in input layer, three middle layers with connections between nodes in different layers and a single value as model target at output layer. The reliability of the proposed structure has been evaluated based on the calculating coefficient of determination, mean square error, and mean absolute percentage error.

5.6.6 Reflection on the State of the Art of Isobaric Heat Capacity of Ionanofluids

The model development cannot proceed without the reliable experimental data. In the case of the isobaric heat capacity of ionanofluids, the effect of nanoparticle loading on their isobaric heat capacity is not clear. Moreover, obtaining reliable experimental isobaric heat capacity data is a challenge, where the selection of appropriate research methodology is crucial. While, the thermal conductivity of ionanofluids has been analyzed, many times and many review papers have been published[118]. Neither a comprehensive critical analysis of the experimental data on the isobaric heat capacity of ionanofluids, nor a thorough evaluation of the existing theoretical models has been undertaken to date. Such a systematic analysis is indispensable and currently constitutes the focus of ongoing research conducted by our group.

5.7 CONCLUSIONS AND OUTLOOK

Recently, many theoretical studies have been carried out to predict the thermal conductivity of ionanofluids. Several models have also been proposed with consideration of various mechanisms. Generally, the theoretical efforts can be categorized into two groups. The first group applied the concept of liquid/solid interfacial nanolayer

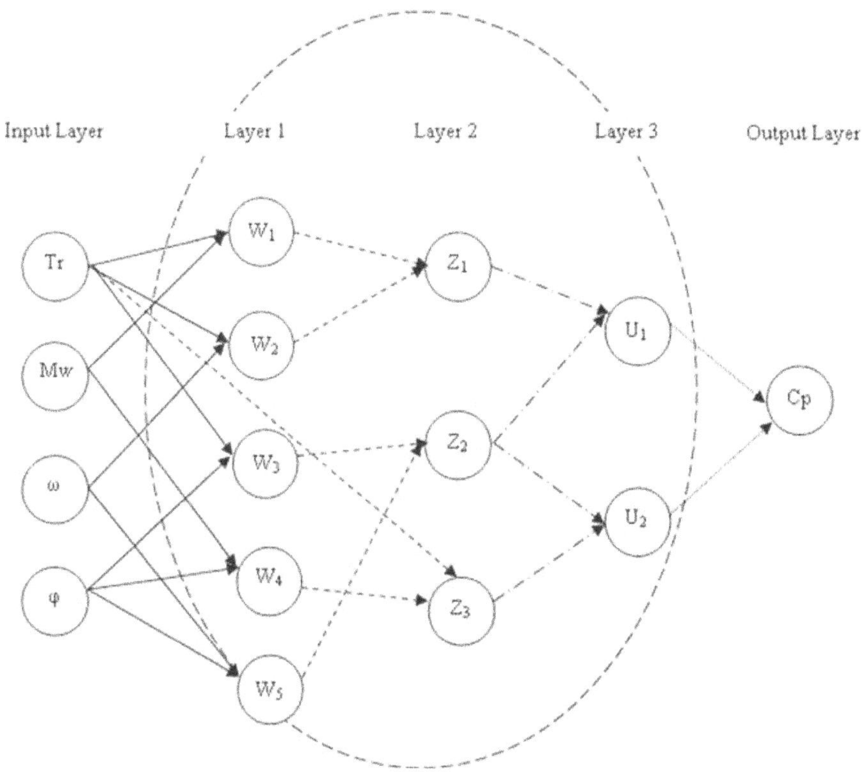

FIGURE 5.33 Structure of developed group method of data handling model with three layers and nodal expressions for each layer. (Reprinted with permission from Sadi[117] Copyright © 2017 Springer Nature.)

to develop models and to explain the improvement of the thermal conductivity in ionanofluids. In most of the cases, the Maxwell, Hamilton–Crosser, or Bruggeman model was modified by considering the nanoparticle to be surrounded by an interfacial layer formed by base liquid. However, surprisingly, an inconsistency was found between the results obtained by the most classical Maxwell and Hamilton–Crosser models. This requires further analysis. The second group of concepts was based on the contribution of dynamic component into thermal conductivity related to particle Brownian motion and aggregation. The importance of Brownian motion–induced nanoconvection in the fluid rather than its direct contribution of the particle motion to thermal behavior was postulated. However, the nanoconvection mechanism, has not been fully explored so far. Recently, the concept of heat transfer by Brownian motion was applied to the prediction of the thermal conductivity of nanofluids, with better results. It is expected that this concept will soon be applied to ionanofluids. In the future, it is important to include the viscosity of the base liquid in theoretical research, because both inactive and active Brownian motion of nanoparticles depends on the base liquid viscosity. In general, the viscosity and particularly the rheological behavior of ionanofluids have received limited attention in the literature.

It nevertheless has a significant impact on heat transfer mechanisms. Therefore, research on the mechanisms of heat transport in ionanofluids is still insufficient, and elucidating the experimental characteristics of nanofluids and measurement conditions is crucial for investigating heat transport in nanofluids.

In the case of the isobaric heat capacity of ionanofluids and pure ionic liquids, the most important issue is reliable measurements and critical analysis of the obtained results.

ACKNOWLEDGMENTS

Marzena Dzida is profoundly indebted to Krzysztof Cwynar MSc, Dr. Anna Kolanowska and Dr. Katarzyna Kaczmarek for critical reading of this book chapter. Marzena Dzida thanks National Science Centre (Poland) for financial support of this work (Grant No. 2017/27/B/ST4/02748)

ABBREVIATIONS

$[(C_2F_5)_3PF_3]^-$	tris(pentafluoroethyl)trifluorophosphate anion
$[(CF_3SO_2)_2N]^-$, $[NTf_2]^-$	bis(trifluoromethylsulfonyl)imide anion
$[C_1C_1im][DMP]$	1,3-dimethylimidazolium dimethylphosphate
$[C_2C_1im][C_2SO_4]$	1-ethyl-3-methylimidazolium ethyl sulfate
$[C_2C_1im][N(CN)_2]$	1-ethyl-3-methylimidazolium dicyanamide
$[C_2C_1im][SCN]$	1-ethyl-3-methylimidazolium thiocyanate
$[C_2C_1im]Cl$	1-ethyl-3-methylimidazolium chloride
$[C_3C_1pyr][NTf_2]$	1-propyl-1-methylpyrrolidinium bis(trifluoromethylsulfonyl)imide
$[C_4C_1C_1im][NTf_2]$	1-butyl-2,3-dimethylimidazolium bis(trifluoromethylsulfonyl)imide
$[C_4C_1im][C(CN)_3]$	1-butyl-3-methylimidazolium tricyanomethanide
$[C_4C_1im][NTf_2]$	1-butyl-3-methylimidazolium bis(trifluoromethylsulfonyl) imide
$[C_4C_1im][PF_6]$	1-butyl-3-methylimidazolium hexafluorophosphate
$[C_4C_1pyr][NTf_2]$	1-butyl-1-methylpyrrolidinium bis(trifluoromethylsulfonyl)imide
$[C_6C_1im][BF_4]$	1-hexyl-3-methylimidazolium tetrafluoroborate
$[C_6C_1im][PF_6]$	1-hexyl-3-methylimidazolium hexafluorophosphate
$[CF_3SO_3]^-$	trifluoromethanesulfonate anion
$[N_4111][NTf_2]$	*N*-butyl-*N*,*N*,*N*-trimethylammonium bis(trifluoromethylsulfonyl)imide
$[P_{14,6,6,6}][HexO]$	trihexyl(tetradecyl)phosphonium hexanoate
$[P_{14,6,6,6}]^+$	trihexyl(tetradecyl)phosphonium cation
ANN	artificial neural network
cryo-TEM	transmission electron cryomicroscopy
CTAB	cetyltrimethylammonium bromide
d	diameter
DLVO	Derjaguin–Landau–Verwey–Overbeck

h	thickness of interfacial nanolayer
IL	ionic liquid
INF	ionanofluid
InL	interfacial nanolayer
l	length
MARD	mean absolute relative deviation
MLY	Murshed, Leong, and Yang
MWCNTs	multi-walled carbon nanotubes
NFs	nanofluids
NPs	nanoparticles
PVP40	poly(*N*-vinylpyrrolidone)40
r	radius of the nanoparticle
RD	relative deviation
SWCNTs	single-walled carbon nanotubes

REFERENCES

1. Warrier, P.; Yuan, Y.; Beck, M. P.; Teja, A. S. Heat Transfer in Nanoparticle Suspensions: Modeling the Thermal Conductivity of Nanofluids. *AIChE J.*, **2010**, *56* (12), 3243–3256. https://doi.org/10.1002/aic.12228.
2. Induranga, A.; Galpaya, C.; Vithanage, V.; Indupama, A.; Maduwantha, K.; Gunawardana, N.; Wijesekara, D.; Amarasinghe, P.; Nilmalgoda, H.; Gunasena, K.; et al. Nanofluids for Heat Transfer: Advances in Thermo-Physical Properties, Theoretical Insights, and Engineering Applications. *Energies*, **2025**, *18* (8), 1935. https://doi.org/10.3390/en18081935.
3. Balandin, A. A. Thermal Properties of Graphene and Nanostructured Carbon Materials. *Nat. Mat.*, **2011**, *10* (8), 569–581. https://doi.org/10.1038/nmat3064.
4. Xie, Y.; Wang, X. Thermal Conductivity of Carbon-Based Nanomaterials: Deep Understanding of the Structural Effects. *Green Carbon*, **2023**, *1* (1), 47–57. https://doi.org/10.1016/j.greenca.2023.08.004.
5. Oster, K.; Hardacre, C.; Jacquemin, J.; Ribeiro, A. P. C.; Elsinawi, A. Thermal Conductivity Enhancement Phenomena in Ionic Liquid-Based Nanofluids (Ionanofluids). *Aust. J. Chem.*, **2019**, *72* (2), 21. https://doi.org/10.1071/CH18116.
6. Chereches, E.I.; Bejan, D.; Ibanescu, C.; Danu, M.; Minea, A.A. Ionanofluids with [C2mim][CH3SO3] ionic liquid and alumina nanoparticles: An experimental study on viscosity, specific heat and electrical conductivity. *Chem. Eng. Sci.*, **2021**, *229*, Article 116140.
7. Oster, K.; Hardacre, C.; Jacquemin, J.; Ribeiro, A. P. C.; Elsinawi, A. Ionic Liquid-Based Nanofluids (Ionanofluids) for Thermal Applications: An Experimental Thermophysical Characterization. *Pure Appl. Chem.*, **2019**, *91* (8), 1309–1340. https://doi.org/10.1515/pac-2018-1114.
8. Kazakov, A.; Magee, J.W; Chirico, R.D.; Paulechka, E.; Diky, V.; Muzny, C.D; Kroenlein, K.; Frenkel, M. "NIST Standard Reference Database 147: NIST Ionic Liquids Database – (ILThermo)", Version 2.0; National Institute of Standards and Technology: Gaithersburg MD, **2017**.
9. Fox, E. B.; Visser, A. E.; Bridges, N. J.; Amoroso, J. W. Thermophysical Properties of Nanoparticle-Enhanced Ionic Liquids (NEILs) Heat-Transfer Fluids. *Energy Fuels*, **2013**, *27* (6), 3385–3393. https://doi.org/10.1021/ef4002617.
10. Nieto de Castro, C. A.; Lourenço, M. J. V.; Ribeiro, A. P. C.; Langa, E.; Vieira, S. I. C.; Goodrich, P.; Hardacre, C. Thermal Properties of Ionic Liquids and IoNanofluids of Imidazolium and Pyrrolidinium Liquids. *J. Chem. Eng. Data*, **2010**, *55* (2), 653–661. https://doi.org/10.1021/je900648p.

11. Paul, T. C.; Morshed, A. K. M. M.; Fox, E. B.; Khan, J. A. Thermal Performance of Al2O3 Nanoparticle Enhanced Ionic Liquids (NEILs) for Concentrated Solar Power (CSP) Applications. *Int. J. Heat Mass Transfer*, **2015**, *85*, 585–594. https://doi.org/10.1016/j.ijheatmasstransfer.2015.01.071.
12. Dzida, M.; Boncel, S.; Jóźwiak, B.; Greer, H. F.; Dulski, M.; Scheller, Ł.; Golba, A.; Flamholc, R.; Dzido, G.; Dziadosz, J.; et al. High-Performance Ionanofluids from Subzipped Carbon Nanotube Networks. *ACS Appl. Mater. Interfaces*, **2022**, *14* (45), 50836–50848. https://doi.org/10.1021/acsami.2c14057.
13. Jóźwiak, B.; Dzido, G.; Zorębski, E.; Kolanowska, A.; Jędrysiak, R.; Dziadosz, J.; Libera, M.; Boncel, S.; Dzida, M. Remarkable Thermal Conductivity Enhancement in Carbon-Based Ionanofluids: Effect of Nanoparticle Morphology. *ACS Appl. Mater. Interfaces*, **2020**, *12* (34), 38113–38123. https://doi.org/10.1021/acsami.0c09752.
14. Wang, B.; Wang, X.; Lou, W.; Hao, J. Gold-Ionic Liquid Nanofluids with Preferably Tribological Properties and Thermal Conductivity. *Nanoscale Res. Lett.*, **2011**, *6* (1), 259. https://doi.org/10.1186/1556-276X-6-259.
15. Paredes, X.; Lourenço, M. J.; Castro, C. N. de; Wakeham, W. Thermal Conductivity of Ionic Liquids and IoNanofluids. Can Molecular Theory Help? *Fluids*, **2021**, *6* (3), 116. https://doi.org/10.3390/fluids6030116.
16. Ribeiro, A. P. C.; Vieira, S. I. C.; Goodrich, P.; Hardacre, C.; Lourenço, M. J. V.; De Castro, C. A. N. Thermal Conductivity of $[C_nMim][(CF_3SO_2)_2N]$ and $[C_4mim][BF_4]$ IoNanofluids with Carbon Nanotubes—Measurement, Theory and Structural Characterization. *J. Nanofluids*, **2013**, *2* (1), 55–62. https://doi.org/10.1166/jon.2013.1035.
17. Cwynar, K.; Kolanowska, A.; Dziadosz, J.; Jóźwiak, B.; Dzido, G.; Greer, H. F.; Boncel, Sławomir; Dzida, M. From Varying Inner Structures of Ionic Liquids towards Extremely Heat Conductive Ionanofluids, under Preparation, **2026**.
18. Canongia Lopes, J. N. A.; Pádua, A. A. H. Nanostructural Organization in Ionic Liquids. *J. Phys. Chem. B*, **2006**, *110* (7), 3330–3335. https://doi.org/10.1021/jp056006y.
19. França, J. M. P.; Lourenço, M. J. V.; Murshed, S. M. S.; Pádua, A. A. H.; Nieto de Castro, C. A. Thermal Conductivity of Ionic Liquids and IoNanofluids and Their Feasibility as Heat Transfer Fluids. *Ind. Eng. Chem. Res.*, **2018**, *57* (18), 6516–6529. https://doi.org/10.1021/acs.iecr.7b04770.
20. Kolanowska, A.; Kuziel, A.; Li, Y.; Jurczyk, S.; Boncel, S. Rieche Formylation of Carbon Nanotubes – One-Step and Versatile Functionalization Route. *RSC Adv.*, **2017**, *7* (81), 51374–51381. https://doi.org/10.1039/C7RA10525H.
21. Jóźwiak, B.; Scheller, Ł.; Greer, H. F.; Cwynar, K.; Urbaniec, K.; Dzido, G.; Dziadosz, J.; Jędrysiak, R.; Kolanowska, A.; Blacha, A.; et al. Water-Like Thermal Conductivity of Ionanofluids Containing High Aspect Ratio Multi-Walled Carbon Nanotubes and 1-Ethyl-3-Methylimidazolium-Based Ionic Liquids with Cyano-Functionalized Anions. *J. Mol. Liq.*, **2023**, *391*, 123329. https://doi.org/10.1016/j.molliq.2023.123329.
22. Scheller, L.; Dziadosz, J.; Jóźwiak, B.; Greer, H. F.; Jędrysiak, R.; Kolanowska, A.; Dzido, G.; Cwynar, K.; Boncel, S.; Dzida, M. On Twin Substantially Improved Thermal Properties and Stability of Pyrrolidinium-Based Ionanofluids with Long Multi-Walled Carbon Nanotubes. *J. Mol. Liq.*, **2023**, *390*, 122971. https://doi.org/10.1016/j.molliq.2023.122971.
23. França, J. M. P.; Vieira, S. I. C.; Lourenço, M. J. V.; Murshed, S. M. S.; Nieto de Castro, C. A. Thermal Conductivity of $[C_4mim][(CF_3SO_2)_2N]$ and $[C_2mim][EtSO_4]$ and Their IoNanofluids with Carbon Nanotubes: Experiment and Theory. *J. Chem. Eng. Data*, **2013**, *58* (2), 467–476. https://doi.org/10.1021/je301183r.
24. Maxwell, J. C. *A Treatise on Electricity and Magnetism*; **1881**, 3rd ed.; Clarendon Press: Oxford, UK, **1891**.
25. Leong, K. C.; Yang, C.; Murshed, S. M. S. A Model for the Thermal Conductivity of Nanofluids – The Effect of Interfacial Layer. *J. Nanopart. Res.*, **2006**, *8* (2), 245–254. https://doi.org/10.1007/s11051-005-9018-9.

26. Murshed, S. M. S.; Leong, K. C.; Yang, C. Thermophysical and Electrokinetic Properties of Nanofluids – A Critical Review. *Appl. Therm. Eng.*, **2008**, *28* (17), 2109–2125. https://doi.org/10.1016/j.applthermaleng.2008.01.005.
27. Murshed, S. M. S.; Leong, K. C.; Yang, C. A Combined Model for the Effective Thermal Conductivity of Nanofluids. *Appl. Therm. Eng.*, **2009**, *29* (11), 2477–2483. https://doi.org/10.1016/j.applthermaleng.2008.12.018.
28. Wang, X.; Xu, X.; Choi, S. U. S. Thermal Conductivity of Nanoparticle – Fluid Mixture. *J. Thermophys. Heat Transfer*, **1999**, *13* (4), 474–480. https://doi.org/10.2514/2.6486.
29. Keblinski, P.; Phillpot, S. R.; Choi, S. U. S.; Eastman, J. A. Mechanisms of Heat Flow in Suspensions of Nano-Sized Particles (Nanofluids). *Int. J. Heat Mass Transfer*, **2002**, *45* (4), 855–863. https://doi.org/10.1016/S0017-9310(01)00175-2.
30. França, J. M. P.; Castro, C. A. N. de; Pádua, A. A. H. Molecular Interactions and Thermal Transport in Ionic Liquids with Carbon Nanomaterials. *Phys. Chem. Chem. Phys.*, **2017**, *19* (26), 17075–17087. https://doi.org/10.1039/C7CP01952A.
31. Wang, B.-X.; Zhou, L.-P.; Peng, X.-F. Surface and Size Effects on the Specific Heat Capacity of Nanoparticles. *Int. J. Thermophys.*, **2006**, *27* (1), 139–151. https://doi.org/10.1007/s10765-006-0022-9.
32. Trisaksri, V.; Wongwises, S. Critical Review of Heat Transfer Characteristics of Nanofluids. *Renewable Sustainable Energy Rev.*, **2007**, *11* (3), 512–523. https://doi.org/10.1016/j.rser.2005.01.010.
33. Yu, W.; France, David M.; Routbort, Jules L.; and Choi, S. U. S. Review and Comparison of Nanofluid Thermal Conductivity and Heat Transfer Enhancements. *Heat Transfer Eng.*, **2008**, *29* (5), 432–460. https://doi.org/10.1080/01457630701850851.
34. Özerinç, S.; Kakaç, S.; Yazıcıoğlu, A. G. Enhanced Thermal Conductivity of Nanofluids: A State-of-the-Art Review. *Microfluid. Nanofluid.*, **2010**, *8* (2), 145–170. https://doi.org/10.1007/s10404-009-0524-4.
35. Murshed, S. M. S.; Nieto de Castro, C. A. Contribution of Brownian Motion in Thermal Conductivity of Nanofluids; Proc. World Congress Eng.: London, U.K., **2011**; pp 1905–1909.
36. Murshed, S. M. S.; de Castro, C. A. N. Predicting the Thermal Conductivity of Nanofluids—Effect of Brownian Motion of Nanoparticles. *J. Nanofluids*, **2012**, *1* (2), 180–185. https://doi.org/10.1166/jon.2012.1024.
37. Angayarkanni, S. A.; Philip, J. Review on Thermal Properties of Nanofluids: Recent Developments. *Adv. Colloid Interface Sci.*, **2015**, *225*, 146–176. https://doi.org/10.1016/j.cis.2015.08.014.
38. Aybar, H. Ş.; Sharifpur, M.; Azizian, M. R.; Mehrabi, M.; Meyer, J. P. A Review of Thermal Conductivity Models for Nanofluids. *Heat Transfer Eng.*, **2015**, *36* (13), 1085–1110. https://doi.org/10.1080/01457632.2015.987586.
39. Ribeiro, A. P. C.; Martins, L. M. D. R. S.; Cristino, A. F. Solid–Liquid Interfaces in Ionanofluids. In *Encyclopedia of Solid-Liquid Interfaces (First Edition)*; Wandelt, K., Bussetti, G., Eds.; Elsevier: Oxford, **2024**; pp 651–660. https://doi.org/10.1016/B978-0-323-85669-0.00148-3.
40. Kotia, A.; Borkakoti, S.; Deval, P.; Ghosh, S. K. Review of Interfacial Layer's Effect on Thermal Conductivity in Nanofluid. *Heat Mass Transfer*, **2017**, *53* (6), 2199–2209. https://doi.org/10.1007/s00231-016-1963-6.
41. Xie, H.; Fujii, M.; Zhang, X. Effect of Interfacial Nanolayer on the Effective Thermal Conductivity of Nanoparticle-Fluid Mixture. *Int. J. Heat Mass Transfer*, **2005**, *48* (14), 2926–2932. https://doi.org/10.1016/j.ijheatmasstransfer.2004.10.040.
42. Murshed, S. M. S.; Nieto de Castro, C. A. Superior Thermal Features of Carbon Nanotubes-Based Nanofluids – A Review. *Renewable Sustainable Energy Rev.*, **2014**, *37*, 155–167. https://doi.org/10.1016/j.rser.2014.05.017.

43. Murshed, S.; Nieto De Castro, C. A. *Thermophysical Properties and Heat Transfer Characteristics of Carbon Nanotubes Dispersed Nanofluids*; NOVA Science Publisher, Inc.: New York, USA: New York, **2014**.
44. Ali, N.; Teixeira, J. A.; Addali, A. A Review on Nanofluids: Fabrication, Stability, and Thermophysical Properties. *J. Nanomater.*, **2018**, *2018* (1), 6978130. https://doi.org/10.1155/2018/6978130.
45. Bastos, A. R. N.; Brites, C. D. S.; Rojas-Gutierrez, P. A.; DeWolf, C.; Ferreira, R. A. S.; Capobianco, J. A.; Carlos, L. D. Thermal Properties of Lipid Bilayers Determined Using Upconversion Nanothermometry. *Adv. Funct. Mat.*, **2019**, *29* (48), 1905474. https://doi.org/10.1002/adfm.201905474.
46. Pensado, A. S.; Malberg, F.; Gomes, M. F. C.; Pádua, A. A. H.; Fernández, J.; Kirchner, B. Interactions and Structure of Ionic Liquids on Graphene and Carbon Nanotubes Surfaces. *RSC Adv.*, **2014**, *4* (35), 18017–18024. https://doi.org/10.1039/C4RA02059F.
47. Atkin, R.; Warr, G. G. Structure in Confined Room-Temperature Ionic Liquids. *J. Phys. Chem. C*, **2007**, *111* (13), 5162–5168. https://doi.org/10.1021/jp067420g.
48. Endres, F.; Borisenko, N.; Abedin, S. Z. E.; Hayes, R.; Atkin, R. The Interface Ionic Liquid(s)/Electrode(s): In situSTM and AFM Measurements. *Faraday Discuss.*, **2011**, *154* (0), 221–233. https://doi.org/10.1039/C1FD00050K.
49. Segura, J. J.; Elbourne, A.; Wanless, E. J.; Warr, G. G.; Voïtchovsky, K.; Atkin, R. Adsorbed and Near Surface Structure of Ionic Liquids at a Solid Interface. *Phys. Chem. Chem. Phys.*, **2013**, *15* (9), 3320–3328. https://doi.org/10.1039/C3CP44163F.
50. Shim, Y.; Kim, H. J. Solvation of Carbon Nanotubes in a Room-Temperature Ionic Liquid. *ACS Nano*, **2009**, *3* (7), 1693–1702. https://doi.org/10.1021/nn900195b.
51. Frolov, A. I.; Kirchner, K.; Kirchner, T.; Fedorov, M. V. Molecular-Scale Insights into the Mechanisms of Ionic Liquids Interactions with Carbon Nanotubes. *Faraday Discuss.*, **2011**, *154* (0), 235–247. https://doi.org/10.1039/C1FD00080B.
52. Mohammadi, M.; Foroutan, M. Mixture of Ionic Liquid and Carbon Nanotubes: Comparative Studies of the Structural Characteristics and Dispersion of the Aggregated Non-Bundled and Bundled Carbon Nanotubes. *Phys. Chem. Chem. Phys.*, **2013**, *15* (7), 2482–2494. https://doi.org/10.1039/C2CP43522E.
53. García, G.; Atilhan, M.; Aparicio, S. Theoretical Study of Renewable Ionic Liquids in the Pure State and with Graphene and Carbon Nanotubes. *J. Phys. Chem. B*, **2015**, *119* (37), 12224–12237. https://doi.org/10.1021/acs.jpcb.5b03809.
54. Zhou, H.; Rouha, M.; Feng, G.; Lee, S. S.; Docherty, H.; Fenter, P.; Cummings, P. T.; Fulvio, P. F.; Dai, S.; McDonough, J.; et al. Nanoscale Perturbations of Room Temperature Ionic Liquid Structure at Charged and Uncharged Interfaces. *ACS Nano*, **2012**, *6* (11), 9818–9827. https://doi.org/10.1021/nn303355b.
55. Wang, J.; Chu, H.; Li, Y. Why Single-Walled Carbon Nanotubes Can Be Dispersed in Imidazolium-Based Ionic Liquids. *ACS Nano*, **2008**, *2* (12), 2540–2546. https://doi.org/10.1021/nn800510g.
56. Konatham, D.; Bui, K.N.D.; Papavassiliou, D.V.; and Striolo, A. Simulation Insights into Thermally Conductive Graphene-Based Nanocomposites. *Mol. Phys.*, **2011**, *109* (1), 97–111. https://doi.org/10.1080/00268976.2010.533707.
57. Kapitza, P. L. Heat Transfer and Superfluidity of Helium II. *Phys. Rev.*, **1941**, *60* (4), 354–355. https://doi.org/10.1103/PhysRev.60.354.
58. Pensado, A. S.; Pádua, A. A. H. Solvation and Stabilization of Metallic Nanoparticles in Ionic Liquids. *Angew. Chem. Int. Ed.*, **2011**, *50* (37), 8683–8687. https://doi.org/10.1002/anie.201103096.
59. Schrekker, H. S.; Gelesky, M. A.; Stracke, M. P.; Schrekker, C. M. L.; Machado, G.; Teixeira, S. R.; Rubim, J. C.; Dupont, J. Disclosure of the Imidazolium Cation Coordination and Stabilization Mode in Ionic Liquid Stabilized Gold(0) Nanoparticles. *J. Colloid Interface Sci.*, **2007**, *316* (1), 189–195. https://doi.org/10.1016/j.jcis.2007.08.018.

60. Jóźwiak, B.; Greer, H. F.; Dzido, G.; Kolanowska, A.; Jędrysiak, R.; Dziadosz, J.; Dzida, M.; Boncel, S. Effect of Ultrasonication Time on Microstructure, Thermal Conductivity, and Viscosity of Ionanofluids with Originally Ultra-Long Multi-Walled Carbon Nanotubes. *Ultrason. Sonochem.*, **2021**, *77*, 105681. https://doi.org/10.1016/j.ultsonch.2021.105681.
61. Clancy, A. J.; Sirisinudomkit, P.; Anthony, D. B.; Thong, A. Z.; Greenfield, J. L.; Singh, M. K. S.; Shaffer, M. S. P. Real-Time Mechanistic Study of Carbon Nanotube Anion Functionalisation through Open Circuit Voltammetry. *Chem. Sci.*, **2019**, *10* (11), 3300–3306. https://doi.org/10.1039/C8SC04970J.
62. Lago-Garrido, S.; Schmidt, D. S.; Martín-Alfonso, M. J.; González-García, L. Multi-Walled Carbon Nanotubes Suspensions as Liquid Conductors: Electrical and Mechanical Network Interplay. *Adv. Electron. Mater.*, **2025**, *11* (11), 2400917. https://doi.org/10.1002/aelm.202400917.
63. Atashrouz, S.; Mozaffarian, M.; Pazuki, G. Modeling the Thermal Conductivity of Ionic Liquids and Ionanofluids Based on a Group Method of Data Handling and Modified Maxwell Model. *Ind. Eng. Chem. Res.*, **2015**, *54* (34), 8600–8610. https://doi.org/10.1021/acs.iecr.5b00932.
64. Turian, R. M.; Sung, D.-J.; Hsu, F.-L. Thermal Conductivity of Granular Coals, Coal-Water Mixtures and Multi-Solid/Liquid Suspensions. *Fuel*, **1991**, *70* (10), 1157–1172. https://doi.org/10.1016/0016-2361(91)90237-5.
65. Everett, D. H. Manual of Symbols and Terminology for Physicochemical Quantities and Units, Appendix II: Definitions, Terminology and Symbols in Colloid and Surface Chemistry. *Pure Appl. Chem.*, **1972**, *31* (4), 577–638.
66. Paul, T. C.; Morshed, Akm. M.; Khan, J. A. Nanoparticle Enhanced Ionic Liquids (NEILS) as Working Fluid for the Next Generation Solar Collector. *Procedia Eng.*, **2013**, *56*, 631–636. https://doi.org/10.1016/j.proeng.2013.03.170.
67. Paul, T. C.; Morshed, A. K. M. M.; Fox, E. B.; Khan, J. A. Experimental Investigation of Natural Convection Heat Transfer of Al2O3 Nanoparticle Enhanced Ionic Liquids (NEILs). *Int. J. Heat Mass Transfer*, **2015**, *83*, 753–761. https://doi.org/10.1016/j.ijheatmasstransfer.2014.12.067.
68. Timofeeva, E. V.; Gavrilov, A. N.; McCloskey, J. M.; Tolmachev, Y. V.; Sprunt, S.; Lopatina, L. M.; Selinger, J. V. Thermal Conductivity and Particle Agglomeration in Alumina Nanofluids: Experiment and Theory. *Phys. Rev. E*, **2007**, *76* (6), 061203. https://doi.org/10.1103/PhysRevE.76.061203.
69. Hamilton, R. L.; Crosser, O. K. Thermal Conductivity of Heterogeneous Two-Component Systems. *Ind. Eng. Chem. Fund.*, **1962**, *1* (3), 187–191. https://doi.org/10.1021/i160003a005.
70. Nieto de Castro, C. A.; Murshed, S. M. S.; Lourenço, M. J. V.; Santos, F. J. V.; Lopes, M. L. M.; França, J. M. P. Enhanced Thermal Conductivity and Specific Heat Capacity of Carbon Nanotubes Ionanofluids. *Int. J. Therm.Sci.*, **2012**, *62*, 34–39. https://doi.org/10.1016/j.ijthermalsci.2012.03.010.
71. Paul, T. C.; Morshed, A. K. M. M.; Fox, E. B.; Khan, J. A. Enhanced Thermophysical Properties of NEILs as Heat Transfer Fluids for Solar Thermal Applications. *Appl. Therm. Eng.*, **2017**, *110*, 1–9. https://doi.org/10.1016/j.applthermaleng.2016.08.004.
72. Asleshirin, S.; Mazaheri, H.; Omidkhah, M. R.; Hassani Joshaghani, A. Investigation of Thermophysical Properties of Io Nanofluids Containing Multi-Walled Carbon Nanotubes and Graphene. *Iran. J. Chem. Chem. Eng.*, **2022**, *41* (2), 380–391. https://doi.org/10.30492/ijcce.2020.121326.4273.
73. Atashrouz, S.; Zarghampour, M.; Abdolrahimi, S.; Pazuki, G.; Nasernejad, B. Estimation of the Viscosity of Ionic Liquids Containing Binary Mixtures Based on the Eyring's Theory and a Modified Gibbs Energy Model. *J. Chem. Eng. Data*, **2014**, *59* (11), 3691–3704. https://doi.org/10.1021/je500572t.

74. Eyring, H. Viscosity, Plasticity, and Diffusion as Examples of Absolute Reaction Rates. *J. Chem. Phys.*, **1936**, *4* (4), 283–291. https://doi.org/10.1063/1.1749836.
75. França, J. M. P.; Reis, F.; Vieira, S. I. C.; Lourenço, M. J. V.; Santos, F. J. V.; Nieto De Castro, C. A.; Pádua, A. A. H. Thermophysical Properties of Ionic Liquid Dicyanamide (DCA) Nanosystems. *J. Chem. Thermodyn.*, **2014**, *79*, 248–257. https://doi.org/10.1016/j.jct.2014.05.008.
76. Wang, F.; Han, L.; Zhang, Z.; Fang, X.; Shi, J.; Ma, W. Surfactant-Free Ionic Liquid-Based Nanofluids with Remarkable Thermal Conductivity Enhancement at Very Low Loading of Graphene. *Nanoscale Res. Lett.*, **2012**, *7* (1), 314. https://doi.org/10.1186/1556-276X-7-314.
77. Tinga, W. R.; Voss, W. A. G.; Blossey, D. F. Generalized Approach to Multiphase Dielectric Mixture Theory. *J. Appl. Phys.*, **1973**, *44* (9), 3897–3902. https://doi.org/10.1063/1.1662868.
78. Oster, K. Ionic Liquid – Based Nanofluids for Thermal Application, A Thesis Submitted to The University of Manchester for the Degree of Doctor of Philosophy in the Faculty of Science and Engineering, School of Chemical Engineering and Analytical Science, Thesis - Oster, Kamil. Ionic liquid-based nanofluids for thermal application. The University of Manchester (United Kingdom), **2018**.
79. Cwynar, K; Dzida, M. Private Communication, University of Silesia in Katowice (Poland), **2025**.
80. Wang, B.; Wang, X.; Lou, W.; Hao, J. Ionic Liquid-Based Stable Nanofluids Containing Gold Nanoparticles. *J. Colloid Interface Sci.*, **2011**, *362* (1), 5–14. https://doi.org/10.1016/j.jcis.2011.06.023.
81. Prasher, R.; Bhattacharya, P.; Phelan, P. E. Thermal Conductivity of Nanoscale Colloidal Solutions (Nanofluids). *Phys. Rev. Lett.*, **2005**, *94* (2), 025901. https://doi.org/10.1103/PhysRevLett.94.025901.
82. Cwynar, K.; Dziadosz, J.; Scheller, Ł.; Zorębski, E.; Jędrysiak, R.; Kolanowska, A.; Blacha, A.; Boncel, S.; Dzida, M. On Isobaric Heat Capacity of Ionanofluids with Carbon Nanotubes – an Experimental Study. *J. Mol. Liq.*, **2023**, 122535. https://doi.org/10.1016/j.molliq.2023.122535.
83. Zhou, S.-Q.; Ni, R. Measurement of the Specific Heat Capacity of Water-Based Al2O3 Nanofluid. *Appl. Phys. Lett.*, **2008**, *92* (9), 093123. https://doi.org/10.1063/1.2890431.
84. Shin, D.; Banerjee, D. Enhancement of Specific Heat Capacity of High-Temperature Silica-Nanofluids Synthesized in Alkali Chloride Salt Eutectics for Solar Thermal-Energy Storage Applications. *Int. J. Heat Mass Transfer*, **2011**, *54* (5–6), 1064–1070. https://doi.org/10.1016/j.ijheatmasstransfer.2010.11.017.
85. Parmar, N.; Bendová, M.; Wagner, Z.; Jacquemin, J. A Study of Changes in the Heat Capacity of Carbon Nanotube-Based Ionanofluids Prepared from a Series of Imidazolium Ionic Liquids. *Phys. Chem. Chem. Phys.*, **2022**, *24* (36), 22181–22190. https://doi.org/10.1039/D2CP02110B.
86. Hentschke, R. On the Specific Heat Capacity Enhancement in Nanofluids. *Nanoscale Res. Lett.*, **2016**, *11* (1), 88. https://doi.org/10.1186/s11671-015-1188-5.
87. Evans, W.; Fish, J.; Keblinski, P. Role of Brownian Motion Hydrodynamics on Nanofluid Thermal Conductivity. *Appl. Phys. Lett.*, **2006**, *88* (9), 093116. https://doi.org/10.1063/1.2179118.
88. Zhou, X. F.; Gao, L. Thermal Conductivity of Nanofluids: Effects of Graded Nanolayers and Mutual Interaction. *J. Appl. Phys.*, **2008**, *103* (8), 083503. https://doi.org/10.1063/1.2904924.
89. Anoop, K. B.; Sundararajan, T.; Das, S. K. Effect of Particle Size on the Convective Heat Transfer in Nanofluid in the Developing Region. *Int. J. Heat Mass Transfer*, **2009**, *52* (9), 2189–2195. https://doi.org/10.1016/j.ijheatmasstransfer.2007.11.063.

90. Jung, J.-Y.; Oh, H.-S.; Kwak, H.-Y. Forced Convective Heat Transfer of Nanofluids in Microchannels. *Int. J. Heat Mass Transfer*, **2009**, *52* (1), 466–472. https://doi.org/10.1016/j.ijheatmasstransfer.2008.03.033.
91. Hai, T.; Basem, A.; Alizadeh, A.; Sharma, K.; Jasim, D. J.; Rajab, H.; Mabrouk, A.; Kolsi, L.; Rajhi, W.; Maleki, H.; et al. Integrating Artificial Neural Networks, Multi-Objective Metaheuristic Optimization, and Multi-Criteria Decision-Making for Improving MXene-Based Ionanofluids Applicable in PV/T Solar Systems. *Sci. Rep.*, **2024**, *14* (1), 29524. https://doi.org/10.1038/s41598-024-81044-3.
92. Parashar, N.; Aslfattahi, N.; Yahya, S. M.; Saidur, R. ANN Modeling of Thermal Conductivity and Viscosity of MXene-Based Aqueous IoNanofluid. *Int. J. Thermophys.*, **2021**, *42* (2), 24. https://doi.org/10.1007/s10765-020-02779-5.
93. Sundar, L. S.; Tavares, S. M. O.; Ramana, E. V.; Pereira, A. M. B. Thermophysical Properties, and Figures-of-Merit Analyses of Nanodiamond/CuO Ionanofluids: Experimental and Artificial Neural Network Predictions. *J. Ionic Liq.*, **2024**, *4* (2), 100113. https://doi.org/10.1016/j.jil.2024.100113.
94. Shaik, N. B.; Inayat, M.; Benjapolakul, W.; Bakthavatchalam, B.; Barewar, S. D.; Asdornwised, W.; Chaitusaney, S. Artificial Neural Network Modeling and Optimization of Thermophysical Behavior of MXene Ionanofluids for Hybrid Solar Photovoltaic and Thermal Systems. *Therm. Sci. Eng. Prog.*, **2022**, *33*, 101391. https://doi.org/10.1016/j.tsep.2022.101391.
95. Ben Hamida, M. B.; Ali, A. B. M.; Sawaran Singh, N. S.; Mostafa, L. Optimization of MXene-Based Aqueous Ionic Liquids for Solar Systems Using Conventional and AI-Based Techniques. *Sci. Rep.*, **2025**, *15* (1), 20565. https://doi.org/10.1038/s41598-025-06702-6.
96. Sundar, L. S. Graphene Oxide Based [EMIM]Cl Ionanofluids in a Tube and Their Heat Transfer, and Friction Factor Analyses under High Prandtl Numbers: Experimental and ANN Predictions. *J. Ionic Liq.*, **2025**, *5* (1), 100144. https://doi.org/10.1016/j.jil.2025.100144.
97. Jasim, D. J.; Rajab, H.; Alizadeh, A.; Sharma, K.; Ahmed, M.; Kassim, M.; AbdulAmeer, S.; Alwan, A. A.; Salahshour, S.; Maleki, H. Enhancing Solar Energy Conversion Efficiency: Thermophysical Property Predicting of MXene/Graphene Hybrid Nanofluids via Bayesian-Optimized Artificial Neural Networks. *Results Eng.*, **2024**, *24*, 102858. https://doi.org/10.1016/j.rineng.2024.102858.
98. Sahebalzamani, S.; Naghizadeh, A.; Mahmoudzadeh, A.; Ghader, S.; Hemmati-Sarapardeh, A. Modeling the Thermophysical Properties of Alumina Nanoparticles Enhanced Ionic Liquids (NEILs) Using Advanced Intelligent Techniques. *Int. J. Hydrog. Energy*, **2025**, *158*, 150375. https://doi.org/10.1016/j.ijhydene.2025.150375.
99. Musiał, M.; Malarz, K.; Mrozek-Wilczkiewicz, A.; Musiol, R.; Zorębski, E.; Dzida, M. Pyrrolidinium-Based Ionic Liquids as Sustainable Media in Heat-Transfer Processes. *ACS Sustainable Chem. Eng.*, **2017**, *5* (11), 11024–11033. https://doi.org/10.1021/acssuschemeng.7b02918.
100. Musiał, M.; Kuczak, M.; Mrozek-Wilczkiewicz, A.; Musiol, R.; Zorębski, E.; Dzida, M. Trisubstituted Imidazolium-Based Ionic Liquids as Innovative Heat Transfer Media in Sustainable Energy Systems. *ACS Sustainable Chem. Eng.*, **2018**, *6* (6), 7960–7968. https://doi.org/10.1021/acssuschemeng.8b01317.
101. Hone, J.; Batlogg, B.; Benes, Z.; Johnson, A. T.; Fischer, J. E. Quantized Phonon Spectrum of Single-Wall Carbon Nanotubes. *Science*, **2000**, *289* (5485), 1730–1733. https://doi.org/10.1126/science.289.5485.1730.
102. Kabo, G. J.; Paulechka, E.; Blokhin, A. V.; Voitkevich, O. V.; Liavitskaya, T.; Kabo, A. G. Thermodynamic Properties and Similarity of Stacked-Cup Multiwall Carbon Nanotubes and Graphite. *J. Chem. Eng. Data*, **2016**, *61* (11), 3849–3857. https://doi.org/10.1021/acs.jced.6b00525.

103. Wiriyasart, S.; Suksusron, P.; Hommalee, C.; Siricharoenpanich, A.; Naphon, P. Heat Transfer Enhancement of Thermoelectric Cooling Module with Nanofluid and Ferrofluid as Base Fluids. *Case Stud. Therm. Eng.*, **2021**, *24*, 100877. https://doi.org/10.1016/j.csite.2021.100877.
104. Hussein, A. M.; Bakar, R. A.; Kadirgama, K. Study of Forced Convection Nanofluid Heat Transfer in the Automotive Cooling System. *Case Stud. Therm. Eng.*, **2014**, *2*, 50–61. https://doi.org/10.1016/j.csite.2013.12.001.
105. Sang, L.; Liu, T. The Enhanced Specific Heat Capacity of Ternary Carbonates Nanofluids with Different Nanoparticles. *Sol. Energy Mater. Sol. Cells*, **2017**, *169*, 297–303. https://doi.org/10.1016/j.solmat.2017.05.032.
106. Wang, L.; Tan, Z.; Meng, S.; Liang, D.; Liu, B. Low Temperature Heat Capacity and Thermal Stability of Nanocrystalline Nickel. *Thermochim. Acta*, **2002**, *386* (1), 23–26. https://doi.org/10.1016/S0040-6031(01)00724-9.
107. Rupp, J.; Birringer, R. Enhanced Specific-Heat-Capacity (c_p) Measurements (150–300K) of Nanometer-Sized Crystalline Materials. *Phys. Rev. B*, **1987**, *36* (15), 7888–7890. https://doi.org/10.1103/PhysRevB.36.7888.
108. Staveley, L. a. K.; Tupman, W. I.; Hart, K. R. Some Thermodynamic Properties of the Systems Benzene+Ethylene Dichloride, Benzene+Carbon Tetrachloride, Acetone+Chloroform, and Acetone+Carbon Disulphide. *Trans. Faraday Soc.*, **1955**, *51* (0), 323–343. https://doi.org/10.1039/TF9555100323.
109. Chase, M. W. NIST-JANAF Themochemical Tables. *J. Phys. Chem. Ref. Data*, **1998**, *Monograph 9*, 1–1951.
110. McCullough, J. P.; Finke, H. L.; Gross, M. E.; Messerly, J. F.; Waddington, G. Low Temperature Calorimetric Studies of Seven 1-Olefins: Effect of Orientation Disorder in the Solid State. *J. Phys. Chem.*, **1957**, *61* (3), 289–301. https://doi.org/10.1021/j150549a007.
111. Fortier, J.-L.; Benson, G. C.; Picker, P. Heat Capacities of Some Organic Liquids Determined with the Picker Flow Calorimeter. *J. Chem. Thermodyn.*, **1976**, *8* (3), 289–299. https://doi.org/10.1016/0021-9614(76)90108-7.
112. Zorębski, E.; Góralski, P. Molar Heat Capacities for (1-Butanol+1,4-Butanediol, 2,3-Butanediol, 1,2-Butanediol, and 2-Methyl-2,4-Pentanediol) as Function of Temperature. *J. Chem. Thermodyn.*, **2007**, *39* (12), 1601–1607. https://doi.org/10.1016/j.jct.2007.04.011.
113. Brzóska, K.; Jóźwiak, B.; Golba, A.; Dzida, M.; Boncel, S. Thermophysical Properties of Nanofluids Composed of Ethylene Glycol and Long Multi-Walled Carbon Nanotubes. *Fluids*, **2020**, *5* (4), 241. https://doi.org/10.3390/fluids5040241.
114. Anouti, M.; Jacquemin, J.; Lemordant, D. Volumetric Properties, Viscosities, and Isobaric Heat Capacities of Imidazolium Octanoate Protic Ionic Liquid in Molecular Solvents. *J. Chem. Eng. Data*, **2010**, *55* (12), 5719–5728. https://doi.org/10.1021/je100671v.
115. Shevelyova, M. P.; Paulechka, Y. U.; Kabo, G. J.; Blokhin, A. V.; Kabo, A. G.; Gubarevich, T. M. Physicochemical Properties of Imidazolium-Based Ionic Nanofluids: Density, Heat Capacity, and Enthalpy of Formation. *J. Phys. Chem. C*, **2013**, *117* (9), 4782–4790. https://doi.org/10.1021/jp3059432.
116. Oster, K.; Hardacre, C.; Jacquemin, J.; Ribeiro, A. P. C.; Elsinawi, A. Understanding the Heat Capacity Enhancement in Ionic Liquid-Based Nanofluids (Ionanofluids). *J. Mol. Liq.*, **2018**, *253*, 326–339. https://doi.org/10.1016/j.molliq.2018.01.025.
117. Sadi, M. Determination of Heat Capacity of Ionic Liquid Based Nanofluids Using Group Method of Data Handling Technique. *Heat Mass Transfer*, **2018**, *54* (1), 49–57. https://doi.org/10.1007/s00231-017-2091-7.
118. Wang, W.; Wu, Z.; Li, B.; Sundén, B. A Review on Molten-Salt-Based and Ionic-Liquid-Based Nanofluids for Medium-to-High Temperature Heat Transfer. *J. Therm. Anal. Calorim.*, **2019**, *136* (3), 1037–1051. https://doi.org/10.1007/s10973-018-7765-y.

6 Prospects and Challenges of Ionanocolloids in Thermal Management and Energy

Aswathy Joseph, S M Sohel Murshed, and Suresh Mathew

6.1 INTRODUCTION

Ionanocolloids (INCs) offer significant promise for next-generation thermal management and energy applications. However, further research is needed to address viscosity, stability, and cost barriers before the call for extensive industrial adoption. The synthesis, characterization, and use of INCs have advanced significantly, but the field is still far from reaching the technical scalability, scientific standardization, and translational readiness needed for practical implementation. A critical review of recent research indicates a number of enduring gaps that prevent their effective application in industrial, environmental, and energy systems. The most important of these is the lack of standardized reporting and characterization procedures, which restricts reproducibility between labs and produces inconsistent thermophysical property data. The absence of reliable multiscale quantitative models that can relate macroscopic transport phenomena like thermal conductivity, viscosity, and electrochemical behavior to nanoscale interfacial structure and ionic layering is equally important.

Furthermore, there are currently few environmental and lifecycle assessments (LCAs) of INCs, which leaves their sustainability profile and ecological impact largely undefined. Few attempts have been made to address cost-per-performance metrics, production scalability, or the viability of system integration in economic analyses and scale-up studies [1, 2]. The majority of studies are limited to short-term static stability tests and are unable to replicate industrially typical dynamic, high-pressure, or high-temperature operating conditions. A coordinated interdisciplinary approach that includes standardized methodologies, predictive multiscale modeling, thorough lifecycle and techno-economic assessments, and scalable process engineering is necessary to close these knowledge gaps. For INCs to progress from laboratory curiosity to commercially viable, sustainable, and high-performance thermal management media, these deficiencies must be addressed. In this concluding chapter, we will address the prospects and challenges in INC research one by one.

DOI: 10.1201/9781003595250-6

6.2 PROSPECTS OF IONANOCOLLOIDS

6.2.1 In Thermal Management

When compared to pure ionic liquids (ILs), the most prominent feature of INCs is their capacity to attain a considerable increase in thermal conductivity. Some studies report up to 200% improvement in thermal conductivity, surpassing the conductivity of conventional aqueous coolants [3]. This enhancement due to certain synergistic effects such as Brownian motion of nanoparticles, interfacial phonon coupling, and formation of thermally conductive percolation networks within these INCs should be investigated more. Besides, the inclusion of carbon-based nanomaterials such as graphene, carbon nanotubes (CNTs), and graphene oxide along with different kinds of IL dispersion medium has been found to be very effective in enhancing the thermal conductivity [4]. The innate non-flammable and thermally stable nature of ILs helps certain INCs to withstand high temperatures over and above 200–300°C without decomposition. This is particularly attractive for enhancing the thermal stability and safety of systems using these materials. The low vapor pressure characteristics of ILs also lessen evaporative loss, facilitating their use in concentrated solar power (CSP), battery thermal management, and hybrid photovoltaic–thermal systems [5].

The INC's good prospects are also due to its application versatility and customizability. Heat exchangers, solar thermal collectors, and microprocessor cooling systems have all demonstrated better performance with INCs. Experimental setups have reported increased temperature differentials and improved energy efficiency [6]. Another significant benefit is that their thermophysical properties may be tuned for specific applications by carefully choosing the types of nanoparticles, concentration, surface functionalization, and IL cations and anions.

6.2.2 Prospects in Energy Applications

The INCs have demonstrated an appealing performance for use in solar energy systems. Compared to water-based systems, significant gains in solar thermal collector efficiency, up to a 37% increase in efficiency, were seen. They are particularly appealing for solar energy harvesting and storage due to their improved optical properties and long-term stability [7]. The presence of nanoparticles improves photon absorption and thermal conductivity while minimizing convective losses in INCs. This facilitates more efficient conversion of solar energy to thermal energy even under fluctuating irradiation. INCs can accomplish heat capacities and storage densities higher than conventional fluids [7]. Therefore, they are also promising candidates for thermal energy storage systems. Strong coulombic interactions and lower vapor pressure created by their ionic structure guarantee stability at high operating temperatures. Furthermore, they are ideal for use in latent or sensible heat storage devices intended for long-term thermal energy retention due to their reversible heat absorption and release capabilities.

In comparison to conventional nanofluids, the use of INCs, such as those based on MXene, in photovoltaic and thermal (PV/T) systems results in increased thermal and electrical efficiencies [8]. Enhancing the performance of MXene-based INCs for hybrid PV/T solar systems by combining artificial neural networks is

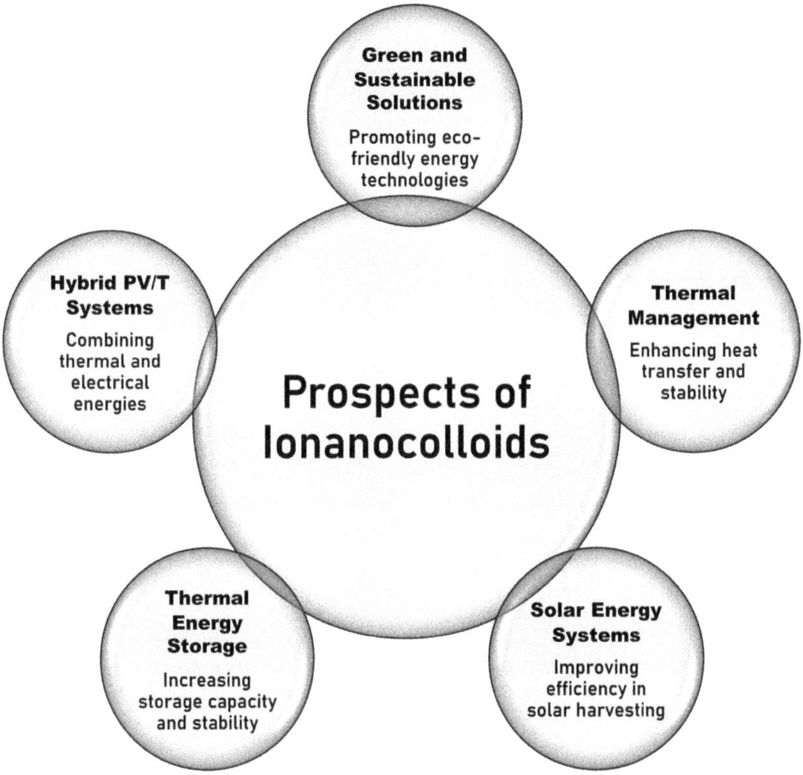

FIGURE 6.1 Scheme shows the prospects of ionanocolloids in thermal management and energy applications.

also increasingly explored [9, 10]. The low volatility of ILs and the photothermal sensitivity of nanoparticles work in concert to improve heat removal from PV cells, lowering operating temperatures and increasing energy conversion. These hybrid systems are particularly promising for decentralized solar technologies, where system dependability and energy efficiency are vital. INCs are seen as environmentally beneficial and support the global movement for greener energy technology. They are environmentally benign due to their negligible vapor pressure, non-flammability, and recyclability [6]. Numerous recent reports demonstrate that bio-based or recyclable precursors can be used to synthesize both ILs and nanoparticles, making it possible to build low-carbon and circular materials for energy infrastructure. Figure 6.1 shows the range of prospects of INCs for various applications, particularly in thermal management and energy applications.

6.3 RESEARCH GAPS IN THE CURRENT LITERATURE

In terms of reaching scientific standardization, technical scalability, and translational readiness, the area is still in its infancy despite notable advancements in the synthesis, characterization, and functional use of INCs. A careful review of the

FIGURE 6.2 Schematic of the main research gap between the INC research and practical applications.

literature reveals a number of enduring research gaps that hinder the practical exploitation of INCs for environmental, industrial, and energy applications [11]. The current research gaps in INC research are shown in Figure 6.2.

6.3.1 Lack of Standardized Characterization and Protocols

One of the most unrelenting obstacles is the lack of standardized protocols for characterizing and reporting INC properties. Minea and Murshed demonstrated a wide scatter in the reported thermal conductivity data of ILs, mostly due to inconsistencies in sample preparation, purity, and measurement method [12]. Fabre and Murshed underline the fact that though extensive studies have been reported on IL thermophysical properties, only a few adopt systematic or standardized methodologies for preparing and characterizing IL–nanoparticle suspensions [13]. Well-known issues include the wide range of methods used to produce INCs, different sonication times, inadequate reporting of aging, types of nanoparticle functionalization, levels of IL purity, and instrument calibration effects [5]. These factors are frequently neglected in discussions and publications. Therefore, standardized experimental and reporting procedures are required to improve the data compatibility, reproducibility, and also, for the development of robust structure–property relationships.

6.3.2 Multiscale Quantitative Models Connecting Transport and Structure

There is a lack of multiscale modeling frameworks that connect macroscopic transport parameters such as those of thermal, rheological, and electrochemical data to interfacial ionic layering and nanoparticle–ion interactions. Current models like Maxwell, Hamilton–Crosser, and Bridgman do not account for ionic microstructure and nanoscale interfacial processes. Hence, they perform poorly for INCs [4]. Density functional theory (DFT) and molecular dynamics (MD) simulations have shown ordered IL layering at the nanoparticle surface with polar and non-polar domains [14]. Yet, efforts to link these microscopic patterns to macroscopic transport coefficients (e.g., viscosity and thermal conductivity) in a predictive manner are limited [15]. The challenge is to bridge the gap between mesoscale or continuum transport models and nanoscale interfacial physics in order to enable prognostic determination of INC features instead of absolutely relying on experiential trial-and-error means.

6.3.3 Evaluation of the Environment and Lifecycle

Although ILs are sometimes referred to as "green" solvents because of their low vapor pressure, it is still difficult to quantify the effects of IL–nanoparticle fluids on the environment and lifecycle. Significant environmental consequences can result from the synthesis of nanoparticles, dispersion chemicals, and the ultimate disposal or recycling of nanofluids. However, there aren't many systematic LCA, carbon footprint evaluations, or end-of-life recycling studies in the INC domain. Claims of "sustainable advanced fluids" might go unsupported in the absence of such data. As a result, the gap here is the absence of a comprehensive environmental or techno-economic assessment of INC systems, encompassing fluid production, deployment, recyclability, and disposal.

6.3.4 Cost Analysis and Scalability

Last but not least, despite the fact that numerous laboratory-scale experiments of INCs exhibit encouraging thermophysical or electrochemical characteristics, very few research deal with cost–benefit analysis, process engineering viability, or scale-up. Reviews point out that the cost of producing IL is still high and that it is difficult to disperse nanoparticles at an industrial scale without sedimentation. The price of ILs is still several orders of magnitude higher than that of conventional fluids, and so far, there has been no thorough cost analysis or study of the lifecycle of INCs, from production to use and disposal. The toxic and environmental impacts of INCs have also gone relatively uninvestigated. Theoretical and predictive models are also in short supply, and any models we do have don't accurately predict the properties of INCs based on the properties of the IL and the nanoparticles it contains. Furthermore, little research has been done on continuous production techniques, economic modeling (e.g., cost per kW, cost per cycle), quality control for large-volume INC batches, or integration with current thermal or electrochemical systems. Even high-performance fluids could find it difficult to make the switch from laboratory to industrial use without these factors. This disparity highlights the necessity of supply-chain analysis, engineering research, and INC manufacturing and deployment economics.

Most of the time, studies on INCs examine the stability of static samples that last for just a few hours or days, but there isn't much research on their behavior under the dynamic conditions that we'd find in real-world applications such as continuous pumping, very high pressure forces, and repeated heating and cooling. Well-known benefits of ILs, such as their high thermal stability, aren't the only thing that's being looked at. Basically no thorough data has been collected on the thermophysical properties of INCs above 200°C, including their viscosity, heat capacity, and thermal conductivity. Heat transfer in INCs is another area that hasn't been given much attention. Besides, researchers have primarily been studying the thermal conductivity, which isn't as important in most heat transfer applications. There's an urgent need for systematic studies on the convection performance of INCs in regular geometries like pipes and channels.

INCs must be filled in these gaps if they are to develop from intriguing academic research into technologically feasible fluids for advanced manufacturing, energy, and environmental systems. The development of validated multiscale models will move the field toward predictive design. The establishment of standard protocols will support reproducible research. The demonstration of robust, cycling-resistant performance will facilitate industrial adoption. Lifecycle and economic analyses will enhance sustainability and cost-competitiveness, and the resolution of scalability issues will facilitate practical implementation. The field of INC has achieved significant progress, but it is currently facing the typical transition phase that many developing materials technologies undergo when going from lab-scale demonstrations to system-scale dependability, economic feasibility, and manufacturability. That shift will happen more quickly if the gaps found are filled.

6.4 MAIN CHALLENGES

The main challenges with INCs are mentioned in Table 6.1. One enduring problem is keeping stable nanoparticle suspensions in ILs. Nanoparticle aggregation or sedimentation can seriously impair electrical and thermal performance. The type, size, surface chemistry, IL structure, and synthesis process of nanoparticles all have a significant impact on stability. Inadequate attention is still paid to long-term colloidal stability under shear or thermal cycling. In general, adding nanoparticles makes ILs more viscous, which impairs flow and convective heat transfer while increasing interfacial energy density [16]. Especially in large-scale systems, this trade-off between decreased fluidity and enhanced thermal conductivity raises the need for pumping power and operating expenses.

ILs and engineered nanoparticles are both expensive, and industrial viability is hampered by the lack of scalable, affordable synthesis methods [17]. The high purity and process control required for repeatable thermophysical properties further restrict the economic viability of INCs.

TABLE 6.1
Summary of the Main Technical and Economic Challenges for INCs

Challenge	Description
Stability	Difficult to maintain long-term colloidal stability; aggregation and sedimentation reduce performance.
High viscosity	Nanoparticle addition increases viscosity, leading to higher pumping power and lower flow efficiency.
Cost and scalability	High material costs and limited scalable synthesis methods restrict industrial feasibility.
Limited understanding	Lack of molecular-level insight and long-term experimental validation.
Pressure drop	Elevated viscosity and density contribute to excessive pressure drop in thermal systems.

In-depth experimental understanding of charge dynamics, interfacial structuring, and interactions between nanoparticles and ILs is still lacking, despite the development of theoretical and computational models [18]. Predicting lifetime stability and thermal reliability is limited by the lack of long-term or practical performance tests [19]. In flow-based systems, high density and viscosity work together to cause pressure drops. This raises the need for more energy to pump and complicates the system as a whole. These hydrodynamic restrictions are particularly important for energy recovery devices and continuous-flow heat exchangers. Overcoming these barriers is essential for their successful industrial application.

6.5 OUTLOOK AND FUTURE PERSPECTIVES

In order to address the above-mentioned gaps, multidisciplinary innovations jointly involving materials design, data science, and sustainability assessment will be needed. The main multidisciplinary innovations that can improve the field of INCs are highlighted in Figure 6.3. One such effort is the recent shift of interest toward "eutectic" ILs, particularly deep eutectic solvents (DES), which is expected to be greener and cheaper base fluids than classic ILs. This probably will improve the economic and environmental profile of INCs since they offer a cost-effective, biodegradable, and tunable alternative to conventional ILs. Another exciting frontier is the concept of "task-specific" ionanofluids (TS-INFs). In these, both the IL and nanoparticle are engineered to carry out specific, optimized functions for receiving, maximizing efficiency, and minimizing unwanted effects. Another successful approach will be to integrate computational modeling and machine learning (ML) to step up the rational design of new IL/NP combinations with predicted thermophysical characteristics and stability prior to experimental synthesis. This will help to reduce experimental cost and time. Conventional INC research mainly focused on single component ILs containing single component NP systems. Therefore, additionally, hybrid INCs that combine more than one type of nanoparticle (e.g., CNT/Al_2O_3 and graphene/SiO_2) are also employable options, as they illustrate synergistic augmentation in conductivity and rheology. In general, as a future outlook, an

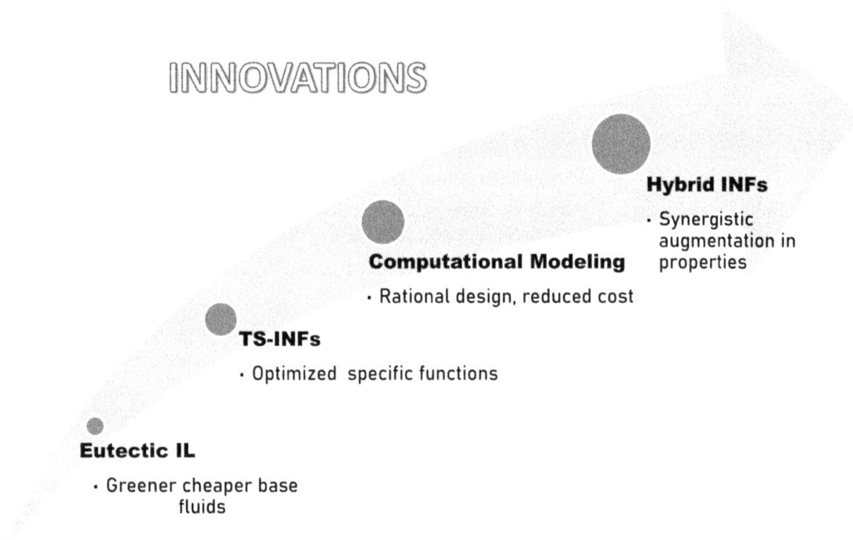

FIGURE 6.3 The multidisciplinary innovations to improve the field of INCs.

integration investigation encompassing the aspects of green chemistry, data-driven materials discovery, and device-oriented testing will be needed to bridge the gap between laboratory and industry. A few important research directions for the future are mentioned in Table 6.2.

6.6 CONCLUSIONS

Right now, the greatest intellectual hurdle is on the mechanistic front: the dispute over the cause of the thermal enhancement, which can be either attributed to the agglomeration of the particles or the creation of nanolayers. Developing predictions from this understanding is also an elusive goal, being heavily reliant on computationally derived

TABLE 6.2
Future Research Directions in INCs for Energy Applications

Research Directions	
Material innovation	Development of new ionic liquids (e.g., amino acid-based, task-specific) and nanoparticle combinations aimed at reducing viscosity while enhancing colloidal stability and thermal performance.
System integration	Optimization of INC and ionanocolloid formulations tailored for specific devices such as solar thermal collectors, electrochemical cells, and heat exchangers to achieve maximum efficiency.
Mechanistic understanding	Advanced molecular-level studies, simulations, and spectroscopic investigations are essential to elucidate nanoparticle–ionic liquid interfacial interactions and guide rational material design.

surface characteristics. As the INC research continues, a top priority is being given to streamlined testing, the use of cutting-edge molecular modeling, and the investigation of new and original base fluid combinations, which will help to close the gap between the lab-bench and the industrial sector. The biodegradability, toxicity, and environmental footprint of INCs need to be thoroughly investigated to gain the trust of the industry and go hand-in-hand with the cost–benefit analysis of the goals we are working toward. On the other hand, the vast potential of INCs means that their applications will continue to multiply and push the limits of their potential.

Future developments could hasten the transition of INCs from lab prototypes to industrial thermal management solutions by combining molecular engineering, ML-driven optimization, and scalable synthesis of inexpensive ILs. To achieve this shift, a multidisciplinary framework connecting process engineering and colloid chemistry will be necessary. Furthermore, the establishment of predictive frameworks for the design of next-generation INCs requires interdisciplinary collaboration that connects chemical engineering, computational thermodynamics, and colloid science. In the age of electrification and renewable energy, INCs may prove to be a key component of sustainable thermal management technologies as our understanding of molecular-level interactions advances.

ABBREVIATIONS

CNTs	carbon nanotubes
CSP	concentrated solar power
DES	deep eutectic solvent
DFT	density functional theory
IL(s)	ionic liquid(s)
INC(s)	ionanocolloid(s)
LCA	lifecycle assessment
MD	molecular dynamics
ML	machine learning
MXene	transition metal carbide/nitride 2D materials
NP(s)	nanoparticle(s)
PV/T	photovoltaic/thermal hybrid system
TS-INC	task-specific INC
wt%	weight percent

REFERENCES

1. Jóźwiak, B., et al., *From lab and up: superior and economic heat transfer performance of ionanofluids containing long carbon nanotubes and 1-ethyl-3-methylimidazolium thiocyanate.* International Journal of Heat and Mass Transfer, 2021. **172**: p. 121161.
2. Fabre, E. and S.S. Murshed. *Critical evaluation of nanofluids and ionanocolloids as heat transfer fluids.* Journal of Physics: Conference Series, 2021. **2116**: p. 012053. https://doi.org/10.1088/1742-6596/2116/1/012053
3. Jóźwiak, B., et al., *Water-like thermal conductivity of ionanofluids containing high aspect ratio multi-walled carbon nanotubes and 1-ethyl-3-methylimidazolium-based ionic liquids with cyano-functionalized anions.* Journal of Molecular Liquids, 2023. **391**: p. 123329.

4. Jóźwiak, B., G. Dzido, and E. Zorębski, *Remarkable thermal conductivity enhancement in carbon-based ionanofluids: effect of nanoparticle morphology.* ACS Appl Mater Interfaces, 2020. **12**(34): p. 38113–38123. https://doi.org/10.1021/acsami.0c09752
5. Fabre, E. and S.S. Murshed, *A review of the thermophysical properties and potential of ionic liquids for thermal applications.* Journal of Materials Chemistry A, 2021. **9**(29): p. 15861–15879.
6. Minea, A.A. and S.M. Sohel Murshed, *Ionic liquids-based nanocolloids— a review of progress and prospects in convective heat transfer applications.* Nanomaterials, 2021. **11**(4). https://doi.org/10.3390/nano11041039
7. Cavieres, J., et al., *Ionanofluids based on ionic liquid mixtures, a new approach as an alternative material for solar energy storage.* Journal of Molecular Liquids, 2022. **351**: p. 118677.
8. Bakthavatchalam, B., et al., *Optimization of thermophysical and rheological properties of Mxene ionanofluids for hybrid solar photovoltaic/thermal systems.* Nanomaterials, 2021. **11**(2): p. 320.
9. Hai, T., et al., *Integrating artificial neural networks, multi-objective metaheuristic optimization, and multi-criteria decision-making for improving MXene-based ionanofluids applicable in PV/T solar systems.* Scientific Reports, 2024. **14**(1): p. 29524.
10. Shaik, N.B., et al., *Artificial neural network modeling and optimization of thermophysical behavior of MXene Ionanofluids for hybrid solar photovoltaic and thermal systems.* Thermal Science and Engineering Progress, 2022. **33**: p. 101391.
11. Pereira, J., R. Souza, and A. Moita, *A review of ionic liquids and their composites with nanoparticles for electrochemical applications.* Inorganics, 2024. **12**(7): p. 186.
12. Minea, A.A. and S.S. Murshed, *A review on development of ionic liquid based nanofluids and their heat transfer behavior.* Renewable and Sustainable Energy Reviews, 2018. **91**: p. 584–599.
13. Fabre, E. and S.S. Murshed, *Thermal Conductivity and Viscosity of Ionanocolloids for Applications in Thermal Energy Systems.* 2021, MHMT.
14. Bernardino, K., *How domain segregation in ionic liquids stabilizes nanoparticles and establishes long-range ordering–a computational study.* ACS Nano, 2024. **18**(31): p. 20389–20400.
15. An, R., N. Wu, Q. Gao, Y. Dong, A. Laaksonen, F.U. Shah, X. Ji, and H. Fuchs, *Integrative studies of ionic liquid interface layers: bridging experiments, theoretical models and simulations.* Nanoscale Horizons, 2024. **9**(4): p. 506–535.
16. Pinto, J., E. Fabre, and S.M.S. Murshed, *Evaluation of stability, viscosity and electrical conductivity of [C2mim][DCA] based ionanocolloids.* Heat and Mass Transfer, 2024. **60**(12): p. 2073–2079.
17. Faizan, M., R. Ahmed, and H.M. Ali, *A critical review on thermophysical and electrochemical properties of Ionanofluids (nanoparticles dispersed in ionic liquids) and their applications.* Journal of the Taiwan Institute of Chemical Engineers, 2021. **124**: p. 391–423.
18. Fabre, E. and S.M.S. Murshed, *A comprehensive review of thermophysical properties and prospects of ionanocolloids in thermal energy applications.* Renewable and Sustainable Energy Reviews, 2021. **151**: p. 111593.
19. Minea, A.A. and S.M. Sohel Murshed, *Ionic liquids-based nanocolloids—a review of progress and prospects in convective heat transfer applications.* Nanomaterials, 2021. **11**(4): p. 1039.

Index

For Product Safety Concerns and Information please contact our EU representative GPSR@taylorandfrancis.com
Taylor & Francis Verlag GmbH, Kaufingerstraße 24, 80331 München, Germany

www.ingramcontent.com/pod-product-compliance
Ingram Content Group UK Ltd.
Pitfield, Milton Keynes, MK11 3LW, UK
UKHW020202200926
475702UK00004B/582

* 9 7 8 1 0 3 2 9 7 7 5 3 9 *